INSTANT ANATOMY

INSTANT ANATOMY

Robert H. Whitaker
MD, MChir, FRCS

Neil R. Borley
MB, BS

Both of the Department of Anatomy
University of Cambridge

FOUR DRAGONS

OXFORD

BLACKWELL SCIENTIFIC PUBLICATIONS
LONDON EDINBURGH BOSTON
MELBOURNE PARIS BERLIN VIENNA

© 1994 by
Blackwell Scientific Publications
Editorial Offices:
Osney Mead, Oxford OX2 0EL
25 John Street, London WC1N 2BL
23 Ainslie Place, Edinburgh EH3 6AJ
238 Main Street, Cambridge
  Massachusetts 02142, USA
54 University Street, Carlton
  Victoria 3053, Australia

Other Editorial Offices:
Librairie Arnette SA
1, rue de Lille
75007 Paris
France

Blackwell Wissenschafts-Verlag GmbH
Kurfürstendamm 57
10707 Berlin
Germany

Blackwell MZV
Feldgasse 13
A-1238 Wien
Austria

First published 1994

Set by Setrite Typesetters, Hong Kong
Printed and bound in Great Britain
at the University Press, Cambridge

DISTRIBUTORS

Marston Book Services Ltd
PO Box 87
Oxford OX2 0DT
(*Orders*: Tel: 0865 791155
        Fax: 0865 791927
        Telex: 837515)

USA
Blackwell Scientific Publications, Inc.
238 Main Street
Cambridge, MA 02142
(*Orders*: Tel: 800 59-6102
        617 876-7000)

Canada
Times Mirror Professional Publishing Ltd
130 Flaska Drive
Markham, Ontario L6G 1B8
(*Orders*: Tel: 800 268-4178
        416 470-6739)

Australia
Blackwell Scientific Publications Pty Ltd
54 University Street
Carlton, Victoria 3053
(*Orders*: Tel: 03 347-5552)

A catalogue record for this title
is available from the British Library

ISBN 0-86542-822-0 (BSP)
ISBN 0-86542-858-1 (Four Dragons)

Library of Congress
Cataloging-in-Publication Data

Whitaker, R.H. (Robert H.)
    Instant anatomy/Robert H. Whitaker,
    Neil R. Borley.
        p.   cm.
    Includes bibliographical references
    and index.
    ISBN 0-86542-822-0
    1. Human anatomy — Outlines, syllabi, etc.
    I. Borley, Neil R. II. Title.
    [DNLM: 1. Anatomy, Regional —
handbooks.
    QS 39 W578i 1994]
    QM31.W55 1994
    611 — dc20

CONTENTS

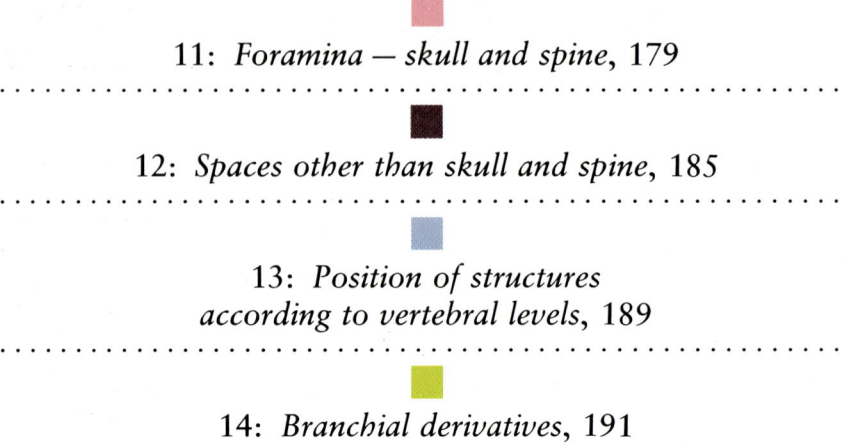

PREFACE

How many times have you looked up the course of an artery or nerve in one of the excellent anatomy textbooks that are available today only to find that the details are spread over several sections of the book and that an instant summary is not available? At times like this you wish there was a quick reference book with all the answers neatly catalogued in dictionary format.

We have attempted to provide such a concise text for rapid reference. Of course, we emphasise that this is not a text for learning anatomy from scratch but one that should be used in conjunction with one of the fuller texts that has stood the test of time. The book is designed for those who already have some working knowledge of anatomy and need to find accurate facts quickly.

Both authors have been sufficiently recent students of anatomy for higher degrees and for teaching undergraduate medical students that each can remember the problems that both students and they themselves encoun-tered. The book has been compiled with this in mind.

It is designed primarily for undergraduate medical students and prospective surgeons who are studying for the first part of the higher degree in surgery. For each of these groups we believe it will be ideal. However, it should also be extremely useful for all clinicians who need to remind themselves of anatomical facts at all stages in their careers and for other professional groups such as nurses, physiotherapists and radiographers.

Inevitably in a book of this size there has been some selection of material for inclusion and no attempt has been made to provide details of minutiae that appear in the fuller texts.

The authors' original artwork was redrawn with a graphics program by Jane Fallows, medical illustrator, to whom the authors owe an immense debt of gratitude for her skill and patience.

ROBERT WHITAKER
NEIL BORLEY
Cambridge, 1994

NOTES ON THE TEXT

The illustrations show the right side of the body as viewed from in front, unless otherwise indicated. The two exceptions are the cervical and brachial plexuses where it makes little difference as to which side they are viewed and they are more conveniently drawn and remembered as they are shown here. Where there might be confusion, a small compass has been added to indicate the left and right *sides of the body.*

Eponymous names appear sparingly and only when they are in common usage. The following abbreviations have been used as appropriately throughout the text.

List of abbreviations

ant	anterior(ly)
art(s)	artery(ies)
br(s)	branch(es)
CMC	carpometacarpal
div(s)	division(s)
ext	external
inf	inferior(ly)
int	internal
IP	interphalangeal
jnt(s)	joint(s)
lat	lateral(ly)
lig(s)	ligament(s)
med	medial(ly)
MC(s)	metacarpal(s)
MCP	metacarpophalangeal
MTP	metatarsophalangeal
MT(s)	metatarsal(s)
N(s)	nerve(s)
post	posterior(ly)
prox	proximal
sup	superior(ly)
TMT	tarsometatarsal

Note: Abbreviations are not used for muscle names or in titles. The following words are always written in full: greater, lesser, middle, superficial and combinations such as mediolateral.

I: ARTERIES

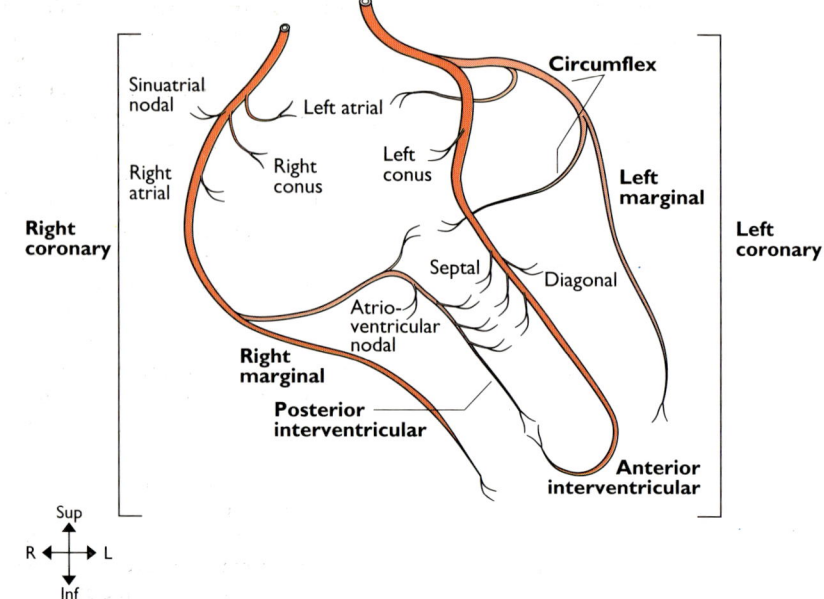

Coronary arteries

CORONARY ARTERIES
From: Ascending aorta
To: Myocardium

Right coronary artery. Originates from the anterior (new nomenclature: right) aortic sinus. It passes anteriorly between the pulmonary trunk and the right auricle to reach the atrioventricular sulcus in which it runs down the anterior surface of the right cardiac border and then onto the inferior surface of the heart. It terminates at the junction of the atrioventricular sulcus and the posterior interventricular groove by anastomosing with the circumflex branch of the left coronary artery and giving off the posterior interventricular (posterior descending) artery. It supplies the right atrium and part of the left atrium, the sinu-atrial node in 60% of cases, the right ventricle, the posterior part of the interventricular septum and the atrio-ventricular node in 80% of cases.

Left coronary artery. Arises from the left posterior (new nomenclature: left) aortic sinus. It passes laterally, posterior to the pulmonary trunk and anterior to the left auricle to reach the atrioventricular groove where it divides into an anterior inter-ventricular (formally left anterior descending) artery and circumflex branches. The circumflex artery runs in the atrio-ventricular sulcus around the left border of the heart to anastomose with the right coronary artery. The anterior inter-ventricular artery descends on the anterior surface of the heart in the anterior interventricular groove and around the apex of the heart into the posterior interven-tricular groove where it anastomoses with the posterior interventricular branch of the right coronary artery. The left coronary artery supplies the left atrium, left ventricle, anterior interventricular septum, sinu-atrial node in 40% of cases and the atrioven-tricular node in 20%.

I

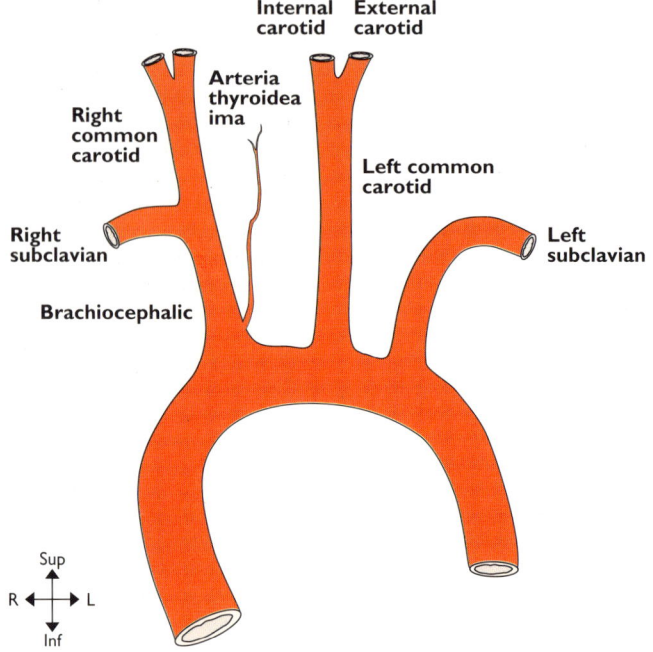

Ascending & arch of aorta

ASCENDING & ARCH OF AORTA
From: **Left ventricle**
To: **Descending aorta**

Ascending aorta. Arises at the vestibule of the left ventricle at the level of the third left costal cartilage and passes upwards and slightly to the left to a point behind the sternum at the level of the manubriosternal joint (second costal cartilage) where it becomes the arch of the aorta. It is enclosed in fibrous and serous pericardium. Anterior to it are the right auricle, the infundibulum of the right ventricle and pulmonary trunk. Posterior, lie the left atrium, the right pulmonary artery and right main bronchus. To the left lie the pulmonary trunk and the left auricle. To the right are the superior vena cava and the right atrium.

Arch of aorta. The arch begins posterior to the manubriosternal joint at the level of the second costal cartilage and passes posterior and to the left, over the left main bronchus to end at the left side of the body of T4 vertebra. Its highest level is the mid point of the manubrium sterni and at this level its three main branches emerge. Anterior and to the left of the arch are (from anterior to posterior) the left phrenic nerve, vagal and sympathetic contributions to the cardiac plexus, and the left vagus. Also, the left superior intercostal vein runs forwards on the arch anterior to the vagus and posterior to the phrenic nerve. Lateral to all these structures are the pleura and left lung. Posterior and to the right of the arch are the trachea, deep cardiac plexus, left recurrent laryngeal nerve, oesophagus, thoracic duct and the body of T4. Inferior to the arch are the pulmonary bifurcation, the left main bronchus, the ligamentum arteriosum and the left recurrent laryngeal nerve. From its superior surface emerge the brachiocephalic artery, the left common carotid and left subclavian arteries. Within the adventitia of the ascending and arch of the aorta lie baro- and chemoreceptors.

Brachiocephalic artery. Arises from the convexity of the aortic arch behind the manubrium sterni and passes upwards and posteriorly to the right. It divides into the right subclavian and right common carotid arteries posterior to the right sterno-clavicular joint. Anterior to it are the left brachiocephalic vein with the right inferior thyroid vein entering it, and the thymic remnants. The artery initially lies anterior to the trachea then passes to lie on its right lateral side. On the right of the artery are the right brachiocephalic vein, upper part of the superior vena cava, the pleura and the cardiac branches of the vagus. The main vagal trunk is more posterolateral. At the origin of the brachiocephalic artery the left common carotid artery lies posteriorly on its left.

continued

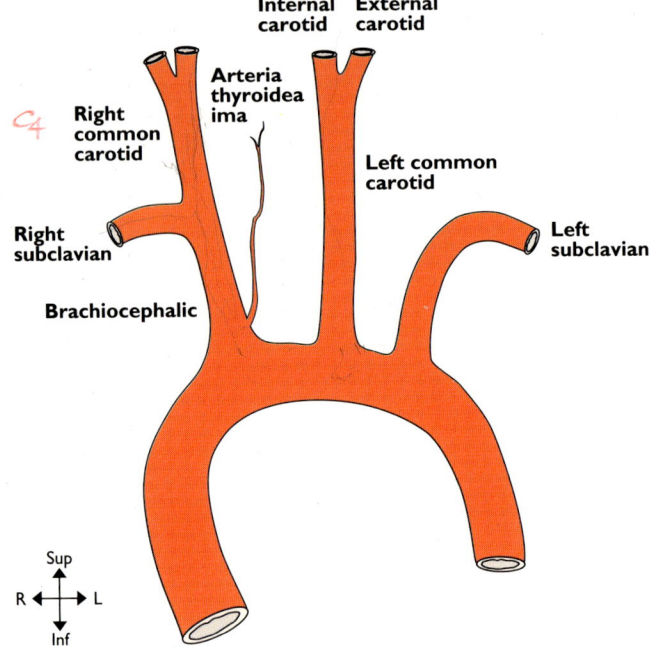

Internal carotid
External carotid
Arteria thyroidea ima
Right common carotid
Left common carotid
Right subclavian
Left subclavian
Brachiocephalic

Sup
R L
Inf

Ascending & arch of aorta

Common carotid arteries. The right common carotid artery arises from the brachiocephalic artery as it divides posterior to the right sternoclavicular joint, whilst the left common carotid arises from the convexity of the aortic arch. Both end as the arteries bifurcate at the level of the upper border of the thyroid cartilage (C4).

Left common carotid artery (thorax). Lying anterior to the thoracic part of this artery are the left brachiocephalic vein and the thymic remnant. Posterior to it in its lower part are the left subclavian artery and the trachea whilst further superiorly there is the left recurrent laryngeal nerve, the thoracic duct and the left side of the oesophagus. On its right at its origin is the brachiocephalic artery but as it ascends the inferior thyroid veins and the trachea come to lie on its right side. To its left lie the vagus, the left phrenic nerve and the left pleura and lung.

Both common carotid arteries (cervical). Ascend in the neck slightly laterally from a point posterior to the sternoclavicular joint to end at the level of the upper border of the thyroid cartilage (C4) at which point there is a dilatation — the carotid sinus (a baroreceptor). On the posterior aspect of the bifurcation there is the carotid body (a chemoreceptor). Lying between left and right arteries, and medial to each, progressively from below are the trachea, recurrent laryngeal nerves, thyroid gland, larynx and pharynx. Each artery lies in its carotid sheath with the internal jugular vein lateral to it and the vagus nerve between and posterior to them both.

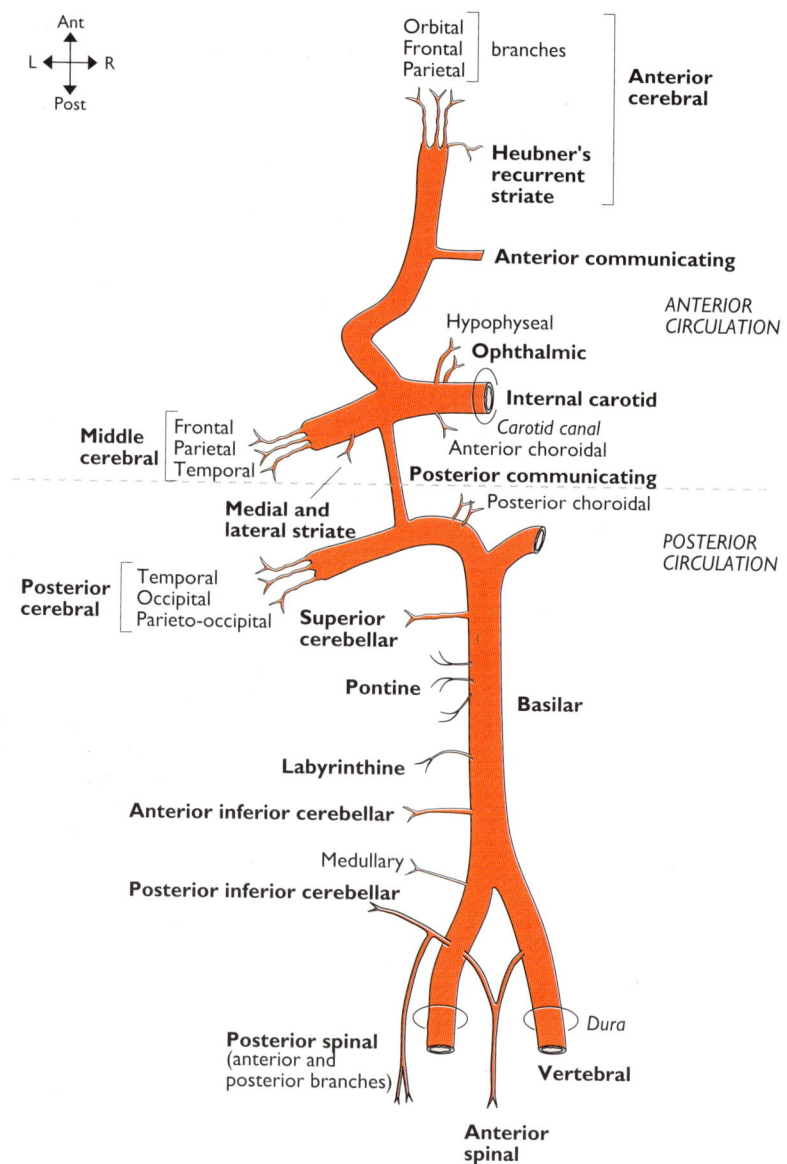

Internal carotid, vertebrobasilar system & circle of Willis
Note: (1) Labyrinthine usually arises from anterior inferior cerebellar; (2) posterior spinal may come from vertebral

INTERNAL CAROTID ARTERY, VERTEBROBASILAR SYSTEM & CIRCLE OF WILLIS

From: Bifurcation of the common carotid
   arts (C4) & first parts of subclavian
   arts

To: Terminal brs

The internal carotid artery angles from the bifurcation slightly posteriorly to reach the carotid canal through which it enters the skull to end as middle and anterior cerebral arteries. At its origin it possesses a dilatation in which lie the carotid sinus and body. In the neck it is crossed laterally by, from below up, the pharyngeal branch of the vagus (X), stylopharyngeus, glossopharyngeal nerve (IX) and styloglossus. It lies on the pharyngeal wall and the pharyngobasilar fascia. Within the carotid canal it turns 90 degrees anteriorly to run through the petrous temporal bone where it lies medial to the middle ear. It then turns 90 degrees superiorly to pass across the upper limit of the foramen lacerum. It then turns 90 degrees anteriorly to pass forwards, lateral to the body of the sphenoid which it grooves. Here it lies in the medial wall of the cavernous sinus with the abducent nerve (VI) on its lateral side. At the anterior end of the cavernous sinus it turns 90 degrees superiorly then 90 degrees posteriorly to pass medial to the anterior clinoid process and lateral to the pituitary stalk and optic chiasma. It ends as terminal branches on the medial surface of the temporal lobe.

Anterior cerebral artery is formed by the bifurcation of the internal carotid artery. It passes anteriorly over the optic nerve to arch over the genu of the corpus callosum on the medial aspect of the cerebral hemispheres where it ends as terminal branches.

Middle cerebral artery is formed by the bifurcation of the internal carotid artery. It runs laterally into the sylvian fissure then posterosuperiorly in the sulcus where it divides into terminal branches.

Basilar artery is formed by the junction of the left and right vertebral arteries (see subclavian artery, pp. 16–19) anterior to the upper medulla. From there it ascends lying angled forwards between the pons and the clivus in a slight depression on the anterior surface of the pons. It terminates at the upper border of the pons as posterior cerebral arteries.

Posterior cerebral artery is formed by the bifurcation of the basilar artery. It passes laterally around the cerebral peduncle to run posteriorly above the tentorium cerebelli on the inferomedial surface of the occipital lobe where it divides into terminal branches.

(Other branches of the internal carotid artery, not illustrated, are caroticotympanic, pterygoid and cavernous arteries.)

I

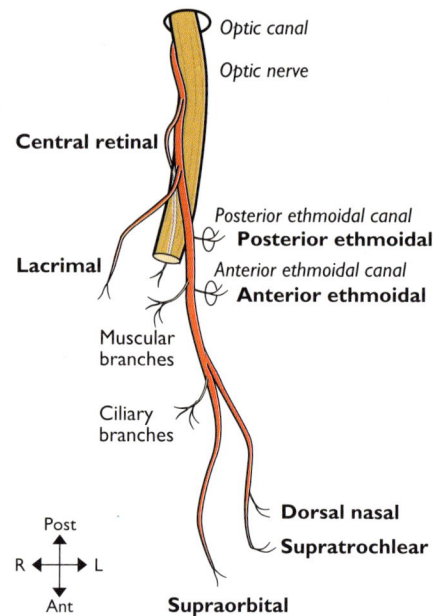

Optic canal

Optic nerve

Central retinal

Posterior ethmoidal canal
Posterior ethmoidal

Lacrimal

Anterior ethmoidal canal
Anterior ethmoidal

Muscular
branches

Ciliary
branches

Post

R ← → L

Ant

Dorsal nasal

Supratrochlear

Supraorbital

Ophthalmic artery
Note: right side viewed from above

OPHTHALMIC ARTERY
From: **Internal carotid art**
To: **Terminal brs in orbit**

It arises from the internal carotid artery as it lies medial to the anterior clinoid process and runs anteriorly through the optic canal where it lies initially inferolateral to the optic nerve, giving off its central retinal branch which runs inferior to the optic nerve. The artery spirals laterally piercing the dural sheath of the nerve and then spirals superiorly directly above the nerve onto its superomedial aspect. It then continues medially between superior oblique and medial rectus to pass out of the cone of muscles to reach the medial wall of the orbit. The artery continues forwards to terminate at the medial orbital border deep to the superior tarsal plate as branches which leave the orbit to anastomose with branches of the facial artery.

(Other branches, not illustrated, (1) of ophthalmic artery are anterior meningeal and medial palpebral arteries; (2) of lacrimal artery are lateral palpebral, zygomatic and recurrent meningeal arteries; (3) of muscular is anterior ciliary artery.)

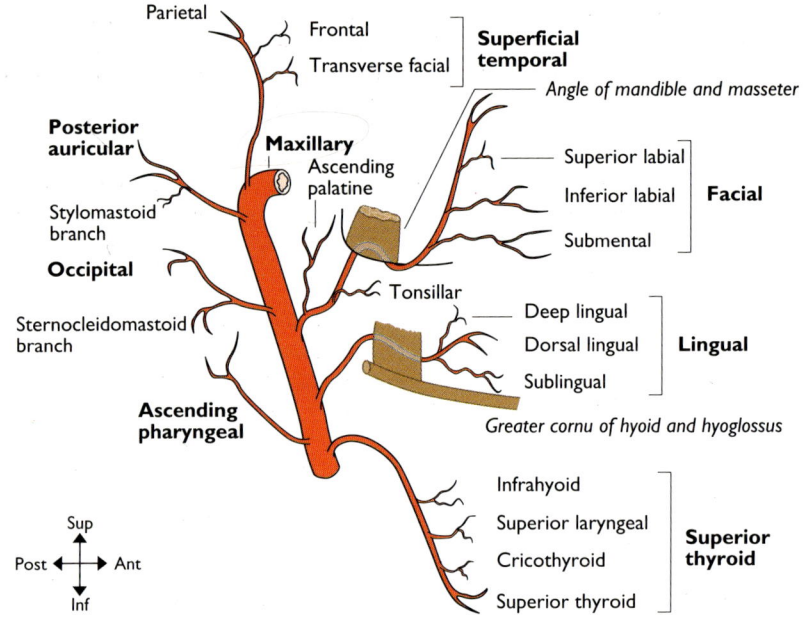

External carotid artery

EXTERNAL CAROTID ARTERY

From: Upper border of thyroid cartilage (C4)

To: Terminal brs within parotid gland post to neck of mandible

The artery arises within the carotid sheath from the bifurcation of the common carotid artery. It lies at first anteromedial to the internal carotid artery but spirals over it to come to lie lateral to it at the level of C2. Initially, it angles slightly forwards then curves backwards as it ascends to enter the parotid gland between deep and superficial lobes. During its course it is crossed by, from below upwards: the upper root of the ansa cervicalis, the hypoglossal nerve, the posterior belly of digastric, stylohyoid, the stylohyoid ligament and the facial nerve (within the parotid). Passing between it and the internal carotid artery are, from below upwards, glossopharyngeal nerve (IX), stylopharyngeus and pharyngeal branch of the vagus (X). It lies on, from below upwards, pharyngeal wall, superior laryngeal branch of the vagus (X) and deep parotid lobe.

Superior thyroid artery. Arises from the anterior surface of the external carotid artery near its origin and runs inferiorly and forwards deep to omohyoid and lateral to the inferior constrictor and external laryngeal nerve to reach the upper pole of the thyroid gland.

Lingual artery. Runs superiorly looping over the greater cornu of the hyoid bone and passes medially (deep) to hyoglossus then into the substance of the tongue.

Facial artery. Arises from the anteromedial surface of the external carotid artery and runs above the hyoid bone deep to digastric and passes upwards to reach the posterior surface of the submandibular gland which it grooves deeply, lying medial to the body of the mandible. Here it lies on superior constrictor, directly lateral to the palatine tonsil. It then follows a tortuous course looping at first inferiorly then upwards around the lower border of the mandible to cross the bone anterior to the insertion of masseter (where it is easily palpable). It then runs in the superficial tissues of the face towards the angle of the mouth where it turns superiorly towards the medial canthus of the eye. (Other branches, not illustrated, are glandular (to submandibular gland) and lateral nasal arteries.)

Superficial temporal artery. Runs superiorly between the deep and superficial lobes of the parotid gland, over the posterior end of the zygomatic process (where it is easily palpable) and terminates in the sub-cutaneous tissues of the lateral scalp.

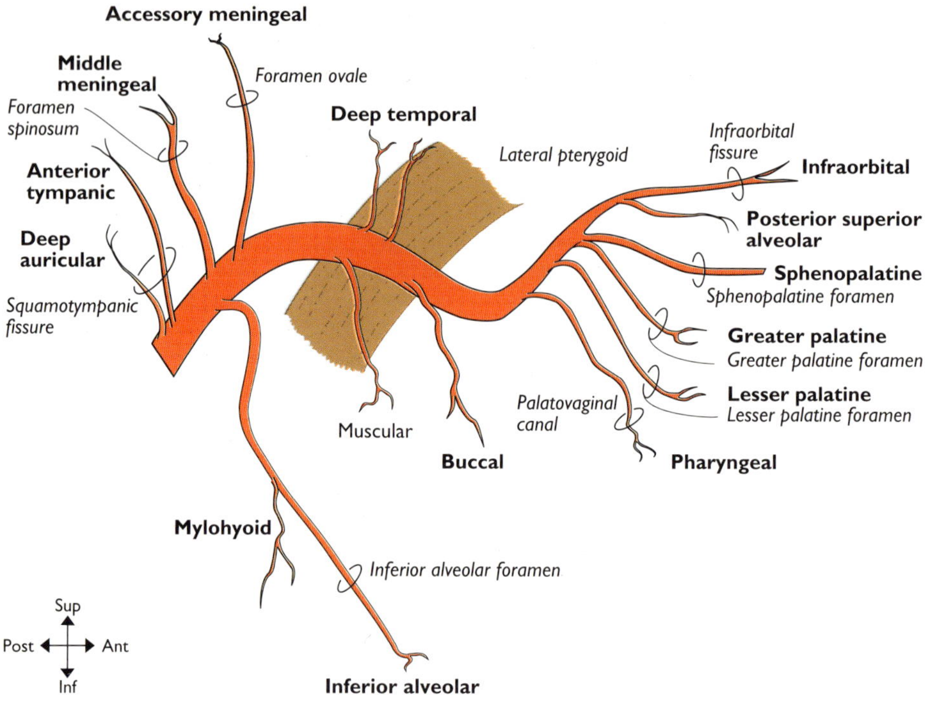

I

Accessory meningeal

**Middle
meningeal**

Foramen ovale

Deep temporal

*Foramen
spinosum*

Lateral pterygoid

*Infraorbital
fissure*

Infraorbital

**Anterior
tympanic**

**Posterior superior
alveolar**

**Deep
auricular**

Sphenopalatine

Sphenopalatine foramen

*Squamotympanic
fissure*

Greater palatine

Greater palatine foramen

Lesser palatine

Lesser palatine foramen

*Palatovaginal
canal*

Muscular

Buccal

Pharyngeal

Mylohyoid

Inferior alveolar foramen

Sup

Post ← → Ant

Inf

Inferior alveolar

Maxillary artery

....

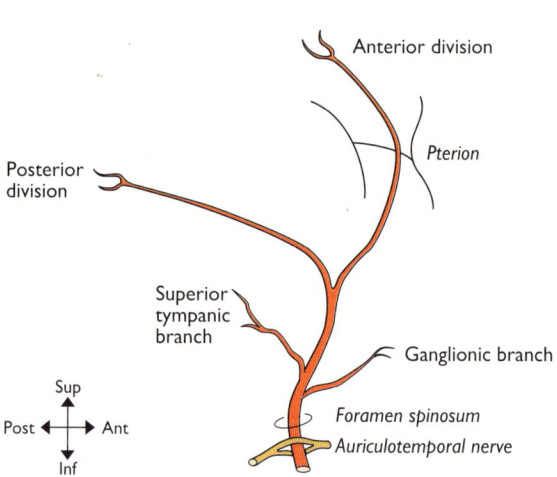

Anterior division

Pterion

Posterior
division

Superior
tympanic
branch

Ganglionic branch

Sup

Post ← → Ant

Inf

Foramen spinosum

Auriculotemporal nerve

Middle meningeal artery

MAXILLARY ARTERY
From: **External carotid within parotid gland**
To: **Terminal brs in pterygopalatine fossa**

It arises from the external carotid artery within the parotid gland posterior to the neck of the mandible and ends as the sphenopalatine artery. The artery is divided into three portions by its relationship posterior, in, or anterior to the lateral pterygoid muscle. The first part passes deep to the neck of the mandible between the bone and the sphenomandibular ligament and runs anteriorly over the inferior alveolar nerve to reach the border of the lateral pterygoid. The second part angles anteromedially between the two heads of lateral pterygoid between anterior and posterior divisions of the mandibular nerve. The third part leaves the lateral pterygoid to enter the pterygopalatine fossa where it terminates as branches which accompany the branches of the maxillary division of the trigeminal nerve (Vb).

Inferior alveolar artery. Passes infero-laterally below the inferior alveolar nerve onto the medial surface of the ramus of the mandible which it grooves as it enters the inferior alveolar (mandibular) foramen in the mandible. It is distributed along the mandibular canal to the lower jaw and teeth. Its terminal branch appears as the mental branch through the mental foramen.

(Other branches, not illustrated, (1) of maxillary artery (third part) is artery of pterygoid canal; (2) of inferior alveolar artery are dental and mental; (3) of infra-orbital artery are dental and anterior superior alveolar; (4) of posterior superior alveolar artery is dental.)

MIDDLE MENINGEAL ARTERY
From: **First part of maxillary art**
To: **Terminal brs**

It arises from the superomedial surface of the maxillary (first part) to run between the two rootlets of the auriculotemporal nerve as it passes vertically into the foramen spinosum in the greater wing of the sphenoid bone. After a very short course laterally over the greater wing of the sphenoid in the middle cranial fossa it divides into anterior and posterior divisions. The anterior division runs anterolaterally on the floor of the middle cranial fossa beneath the dura mater and grooves the greater wing of the sphenoid as it passes upwards to the junction of the lesser and greater wings. Here it may groove deeply or tunnel through the bone at the apex of the greater wing. It passes across the inner aspect of the pterion onto the parietal bone. The posterior division runs almost horizontally posterolateral over the inner aspect of the squamous temporal bone onto the lower parietal bone where it gives terminal branches.

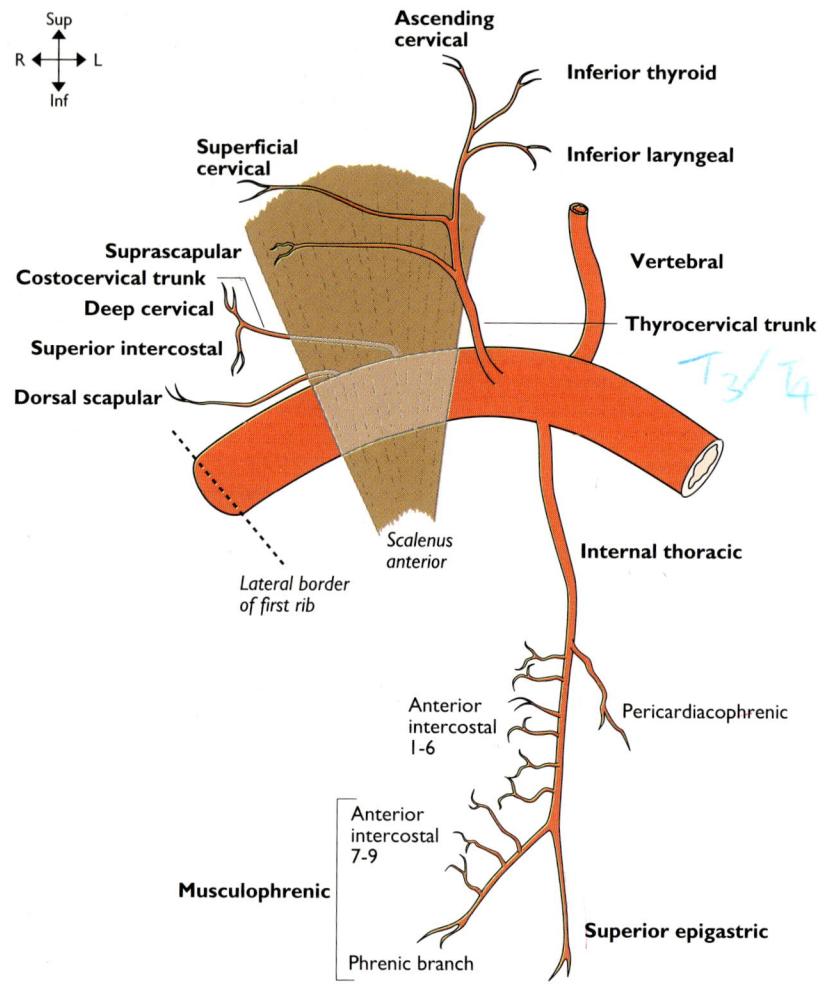

Subclavian artery

Note: (1) The superficial cervical artery is named 'transverse cervical artery' if it gives origin to the dorsal scapular artery instead of the latter arising separately from the second part of the subclavian artery; (2) phrenic branch of musculophrenic artery anastomoses with inferior phrenic artery

SUBCLAVIAN ARTERY
From: Right — brachiocephalic trunk
   Left — aortic arch
To: Axillary art

The subclavian arteries arise as indicated above and end at the outer border of the first rib where they become the axillary arteries. They each have three parts: (1) medial (three branches); (2) behind (two branches); and (3) lateral (no branches) to scalenus anterior.

Right subclavian artery — first part. Arises from the brachiocephalic artery behind the right sternoclavicular joint, lying initially posterior to the right common carotid artery, then passing upwards and laterally to reach the medial side of scalenus anterior. Anterior to this first part are the vagus (X), its cardiac branches, sympathetic nerves, the internal jugular and vertebral veins. The ansa subclavia (sympathetic nerves) curls around the artery to lie both anterior and posterior to it. As the artery arches laterally the suprapleural membrane and the right recurrent laryngeal nerve lie inferior and posterior to it.

Left subclavian artery — first part. Arises from the arch of the aorta just posterior and slightly to the left of the origin of the left common carotid artery at the level of the intervertebral disc of T3/T4. It passes upwards and then, behind the left sterno-clavicular joint, it arches laterally over the suprapleural membrane to the medial edge of scalenus anterior. Anterior to it in the thorax are the left common carotid artery, the left brachiocephalic vein, the left vagus and its cardiac branches and the left phrenic nerve. Posterior to it lie the left side of the oesophagus, the thoracic duct and longus colli. Medial to it is the trachea, the left recurrent laryngeal nerve and, more superiorly, the thoracic duct. In the neck it is crossed anteriorly by the left phrenic nerve and the thoracic duct.

Subclavian artery — second part. Lies posterior to scalenus anterior and anterior to scalenus medius. Anterior to scalenus anterior are the phrenic nerve and, slightly inferior, the subclavian vein. Postero-inferior are the suprapleural membrane and the lower trunk of the brachial plexus. Superior to it are the upper and middle trunks of the brachial plexus.

Subclavian artery — third part. Begins at the lateral margin of scalenus anterior and extends to the outer (lateral) margin of the first rib where it becomes the axillary artery. Anterior to it is the external jugular vein and its tributaries. Antero-inferior is the sub-clavian vein. Postero-inferior is the lower trunk of the brachial plexus and the first rib. Posterosuperior are the upper and middle trunks of the brachial plexus.

Vertebral artery (see also internal carotid, vertebrobasilar system & circle of Willis, pp. 8−9). Arises from the posterosuperior aspect of the first part of the subclavian artery and ends where the arteries from the two sides join to form the basilar artery at the lower pons. It angles posteriorly between the medial border of scalenus medius and the lateral border of longus colli in the apex of the pyramidal space before entering the foramen in the transverse process of C6 behind its anterior tubercle (carotid tubercle of Chassaignac). In this first part, the vessel is related anteriorly to the common carotid artery, the vertebral vein, and, more medially, the inferior thyroid artery and middle cervical ganglion. On the left the thoracic duct crosses it anteriorly. Posterior to it are the anterior primary rami of C7 and C8 nerves and more medially the inferior cervical (stellate) ganglion. The second part of the artery ascends within the foramina of the transverse processes of C6 to C1, accompanied by sympathetic nerves and vertebral veins. It passes out posteriorly behind the lateral mass of the atlas before

continued

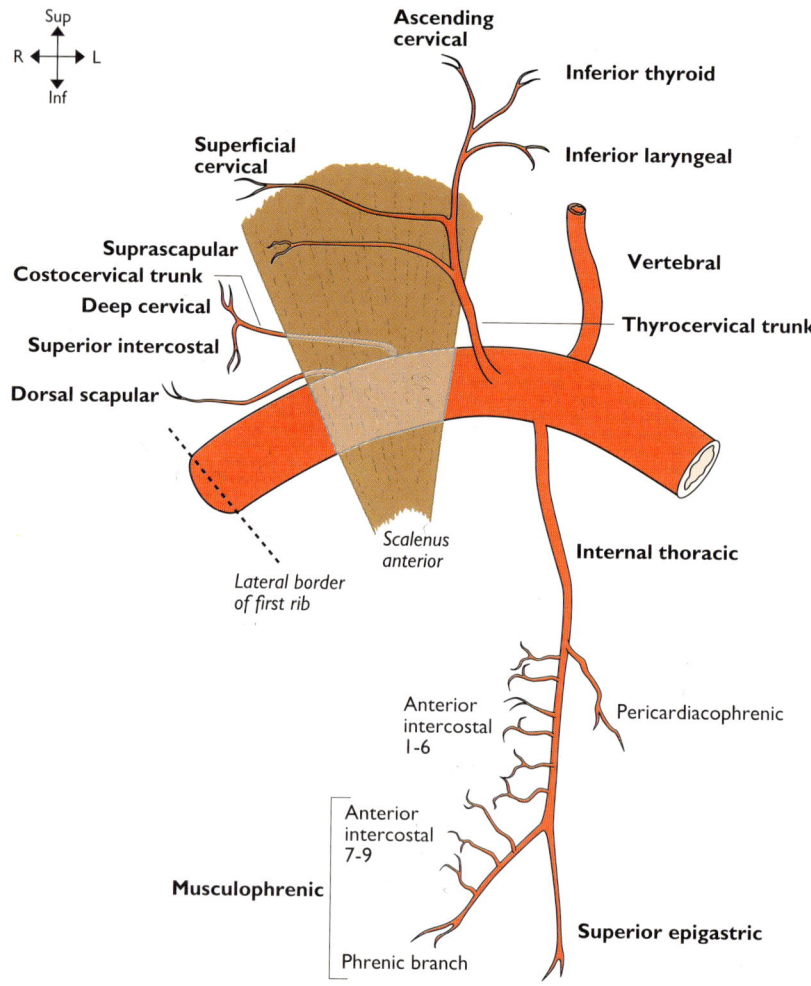

Subclavian artery

Note: (1) The superficial cervical artery is named 'transverse cervical artery' if it gives origin to the dorsal scapular artery instead of the latter arising separately from the second part of the subclavian artery; (2) phrenic branch of musculophrenic artery anastomoses with inferior phrenic artery

turning medially over its posterior arch. It then turns anteriorly to pierce the atlanto-occipital membrane lateral to the cervico-medullary junction. It pierces the dura and arachnoid to ascend supero-medially around the anterior aspect of the medulla where it joins the artery from the opposite side at the lower border of the pons to form the basilar artery. (Other branches, not illustrated, are spinal, meningeal and muscular.)

Internal thoracic artery. Arises from the anterior aspect of the first part of the subclavian artery and passes inferiorly behind the brachiocephalic vein and the phrenic nerve to reach the dome of the pleura. It then angles medially to lie posterior to the upper six costal cartilages, between the internal intercostal and transversus thoracis muscles. It terminates at the 6th intercostal space to give the superior epigastric and musculophrenic arteries. (Other branches, not illustrated, are mediastinal, thymic, sternal and perforating (mammary). Other branches of inferior thyroid artery, not illustrated, are glandular, pharyngeal, oesophageal and tracheal.)

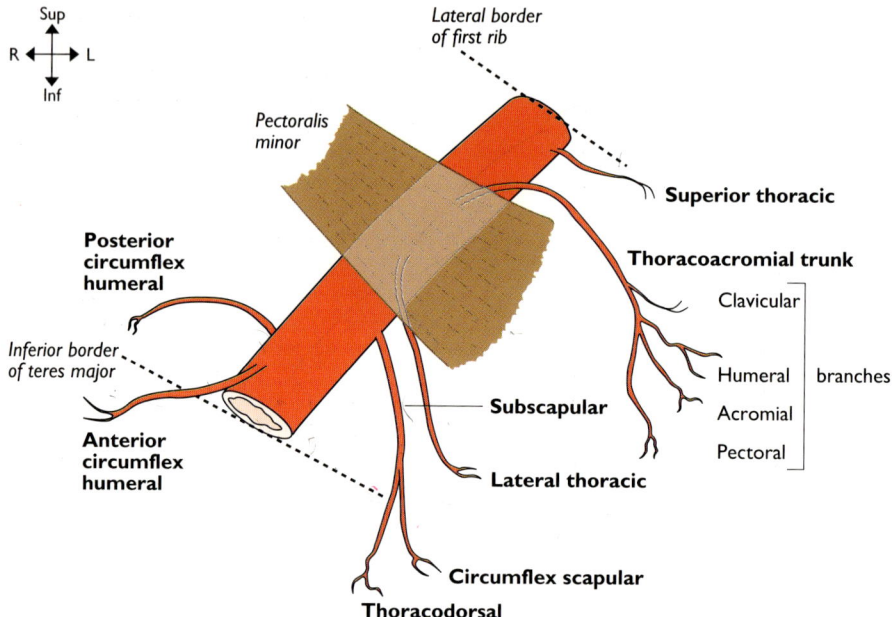

Axillary artery

1. Superior thoracic.

2. Lateral thoracic
   Thoraco-acromial. ← Acromial br
                  Clavicular br
                  Humeral br
                  Pectoral br

3. Subscapular — thoracodorsal
               — circumflex scapula

   Anterior circumflex humeral
   Posterior circumflex humeral

AXILLARY ARTERY
From: **Subclavian art**
To: **Brachial art**

This is the continuation of the subclavian artery. It commences at the lateral border of the 1st rib and ends at the inferior border of teres major to become the brachial artery. It is divided into three parts by pectoralis minor. It is invested in a fascial sheath arising from the prevertebral fascia.

First part is medial to the upper border of pectoralis minor and has one branch. Anterior to it is the clavipectoral fascia, subclavius and the lateral pectoral nerve. The axillary vein is medial whilst posterior to it are the upper part of serratus anterior, the long thoracic nerve, the medial pectoral nerve and the medial cord of the brachial plexus. Lateral to it are the lateral and posterior cords of the brachial plexus.

Second part has pectoralis minor lying anterior to it and has two branches. Medial to it is the axillary vein and medial cord of the brachial plexus. Posterior to it are the posterior cord and subscapularis whilst lateral to it is the lateral cord of the brachial plexus.

Third part extends from the lower border of pectoralis minor to the inferior border of teres major and has three branches. Anterior to it are pectoralis major, the clavipectoral fascia and the median nerve. Medial to it lie the axillary vein and the ulnar nerve. Posterior to it are the radial nerve, teres major, subscapularis and the tendon of latissimus dorsi. On its lateral side lie the musculocutaneous nerve, lateral root (head) of the median nerve, the tendon of biceps in the bicipital groove and coracobrachialis.

I

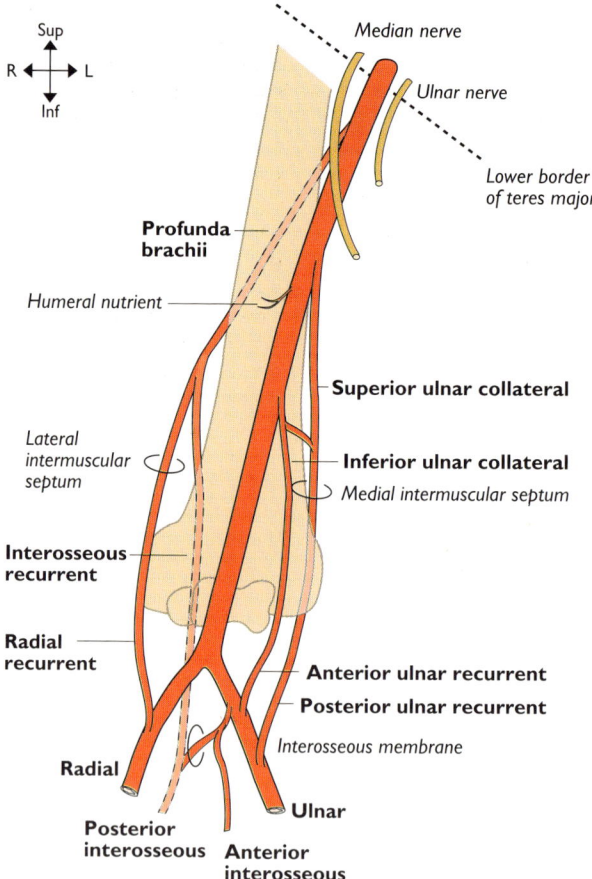

Sup
R ←–→ L
Inf

Median nerve

Ulnar nerve

Lower border
of teres major

**Profunda
brachii**

Humeral nutrient

Superior ulnar collateral

Lateral
intermuscular
septum

Inferior ulnar collateral

Medial intermuscular septum

**Interosseous
recurrent**

**Radial
recurrent**

Anterior ulnar recurrent

Posterior ulnar recurrent

Interosseous membrane

Radial

Ulnar

**Posterior
interosseous**   **Anterior
interosseous**

Brachial artery

BRACHIAL ARTERY
From: **Axillary art**
To: **Radial & ulnar arts**

This is the continuation of the axillary artery beginning at the lower margin of the teres major and ending in the cubital fossa at the level of the neck of the radius as the radial and ulnar arteries. At first it lies medial to the humerus then it spirals around to lie anterior to it. It is superficial throughout its course and accompanied by venae commitantes. It is crossed from lateral to medial by the median nerve in the mid arm and by the bicipital aponeurosis in the cubital fossa. Medial to it is the ulnar nerve in the upper arm and, distally, the median nerve. Lateral to it high up are the median and musculocutaneous nerves. Coraco-brachialis, biceps and its tendon also lie on its lateral side. The artery lies first on the long and then the medial head of triceps, then brachialis in the lower third of the arm.

Arteria profunda brachii. Leaves the _lower aspect_ posteromedial aspect of the brachial artery just below teres major and passes posteriorly between the long and medial heads of triceps with the radial nerve and into the radial groove before breaking up into its terminal branches.

23

I

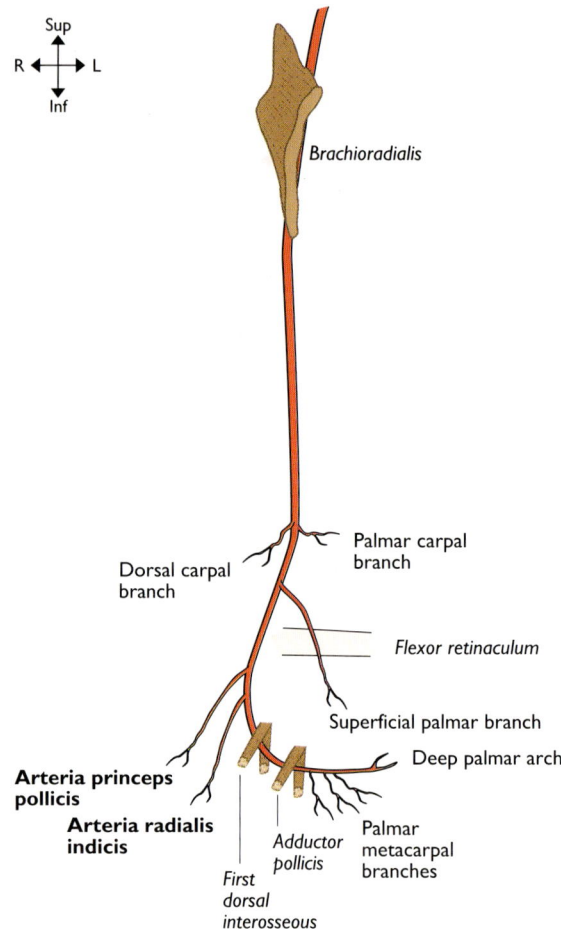

Brachioradialis

Palmar carpal
branch

Dorsal carpal
branch

Flexor retinaculum

Superficial palmar branch

Deep palmar arch

**Arteria princeps
pollicis**

**Arteria radialis
indicis**

*Adductor
pollicis*

Palmar
metacarpal
branches

*First
dorsal
interosseous*

Radial artery

I

RADIAL ARTERY
From: Brachial art in midline of cubital fossa
To: Deep palmar arch in hand

The radial artery arises at the terminal bifurcation of the brachial artery in the cubital fossa at the level of the neck of the radius. It crosses anterior to the biceps tendon to lie initially on supinator. It then passes down the radial side of the forearm lying consecutively on pronator teres, the radial head of flexor digitorum superficialis, flexor pollicis longus and the insertion of pronator quadratus before passing onto the lower end of the radius where its pulse is palpable as it lies lateral to the tendon of flexor carpi radialis. It thus lies deep to brachioradialis and, to a lesser extent, flexor carpi radialis. The superficial branch of the radial nerve lies lateral to it in the forearm. It gives off a palmar carpal branch which contributes to the palmar carpal arch. It then gives off a superficial palmar branch (palmar cutaneous branch) which supplies the thenar muscles before anastomosing with the superficial palmar arch. The radial artery then passes beneath the tendons of abductor pollicis longus and extensor pollicis brevis to enter the anatomical snuff box. It passes across the snuff box on the scaphoid and trapezium and under the tendon of extensor pollicis longus. It gives off a dorsal carpal branch to the dorsal carpal arch which in turn supplies the wrist joint, the dorsal aspects of the metacarpals and the dorsal digital arteries. The radial artery then gives off two named vessels — arteria radialis indicis and princeps pollicis (first palmar metacarpal artery). It next passes down between the two heads of the first dorsal interosseous then between the two heads of adductor pollicis to enter the palm of the hand and form the deep palmar arch. The deep palmar arch lies 1 cm proximal to the superficial palmar arch (ulnar artery). It supplies the palmar metacarpals, gives off a recurrent branch to the palmar carpal arch and three perforating branches which anastomose with the dorsal metacarpal arteries.

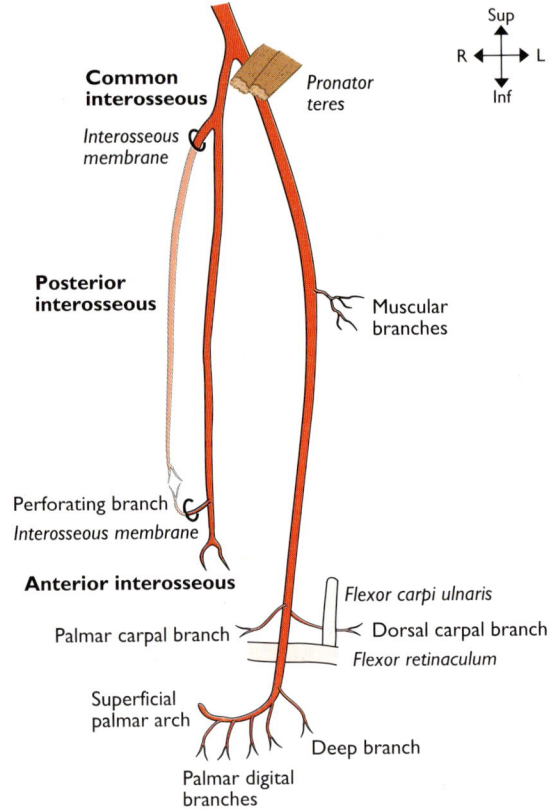

**Common
interosseous**

*Interosseous
membrane*

Pronator
teres

Sup

R ← → L

Inf

**Posterior
interosseous**

Muscular
branches

Perforating branch
Interosseous membrane

Anterior interosseous

Flexor carpi ulnaris

Palmar carpal branch

Dorsal carpal branch

Flexor retinaculum

Superficial
palmar arch

Deep branch

Palmar digital
branches

Ulnar artery

ULNAR ARTERY
From: **Brachial art**
To: **Superficial palmar arch in hand**

The artery arises as the terminal bifurcation of the brachial artery in the cubital fossa at the level of the neck of the radius. It leaves the fossa deep to the deep head of pronator teres and deep to the fibrous arch of flexor digitorum superficialis just lateral to the median nerve to cross beneath the nerve before running down the ulnar side of the forearm. It lies on flexor digitorum profundus with the ulnar nerve on its medial side. It lies lateral to flexor carpi ulnaris before passing superficial to the flexor retinaculum. The dorsal and palmar carpal arteries contribute, with similarly named arteries from the radial artery, to the dorsal and palmar carpal arches. The ulnar artery then gives off a deep branch to the deep palmar arch before forming the superficial palmar arch at the level of the distal border of the extended thumb. The superficial arch supplies the hypothenar eminence and gives off the palmar digital arteries. At the level of pronator teres the ulnar artery gives off the common interosseous artery which divides into anterior and posterior interosseous arteries.

Anterior interosseous artery. Descends on the anterior surface of the interosseous membrane together with the anterior interosseous branch of the median nerve lying between flexor digitorum profundus medially and flexor pollicis longus laterally. Branches perforate the membrane to supply the extensor muscles. Above pronator quadratus it gives off a small branch which descends deep to the muscle to join the palmar carpal arch, then the anterior interosseous artery itself passes posteriorly through the membrane to anastomose with the posterior interosseous artery which descends to join the dorsal carpal arch.

Posterior interosseous artery. Passes back above the interosseous membrane and then runs between supinator superficially and abductor pollicis longus deeply with the deep branch of the radial nerve to descend to supply the extensor muscles of the forearm. It anastomoses with the distal branches of the anterior interosseous artery and dorsal carpal arch.

I

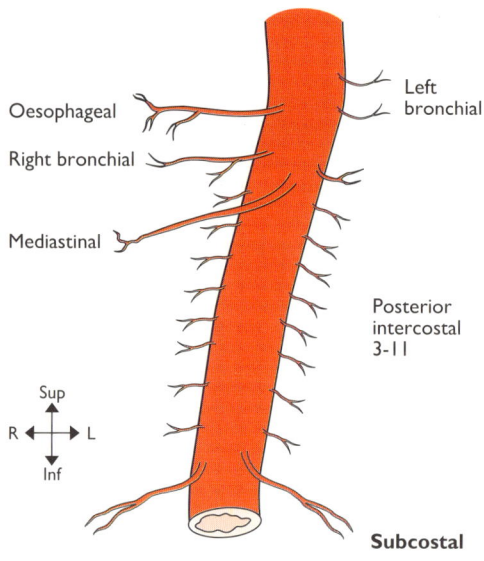

Oesophageal

Left
bronchial

Right bronchial

Mediastinal

Posterior
intercostal
3-11

Sup

R ←→ L

Inf

Subcostal

Thoracic (descending) aorta

THORACIC (DESCENDING) AORTA
From: Arch of aorta
To: Abdominal aorta

This arises as the continuation of the arch of the aorta commencing to the left of the body of T4 and ends as it passes into the abdomen at T12. It grooves the left side of the bodies of T4–T6 vertebrae then it inclines medially to lie in the midline over the lower thoracic vertebrae. It passes out of the thorax at T12 beneath the median arcuate ligament of the diaphragm to become the abdominal aorta. Lying anterior to it from above down are the hilum of the left lung (particularly the left main bronchus), pericardium, left atrium, oesophagus and diaphragm. Posterior lie the necks of the ribs of T5–T6 and the sympathetic chain at that level, the vertebral bodies and hemiazygos veins. To its right lie the right pleura and lung and thoracic duct. The oesophagus and its surrounding plexus of nerves is initially to its right but lower down it crosses the aorta to lie anterior then slightly to the left. To its left are the left pleura and lung. (Other branch, not illustrated, is pericardial.)

I

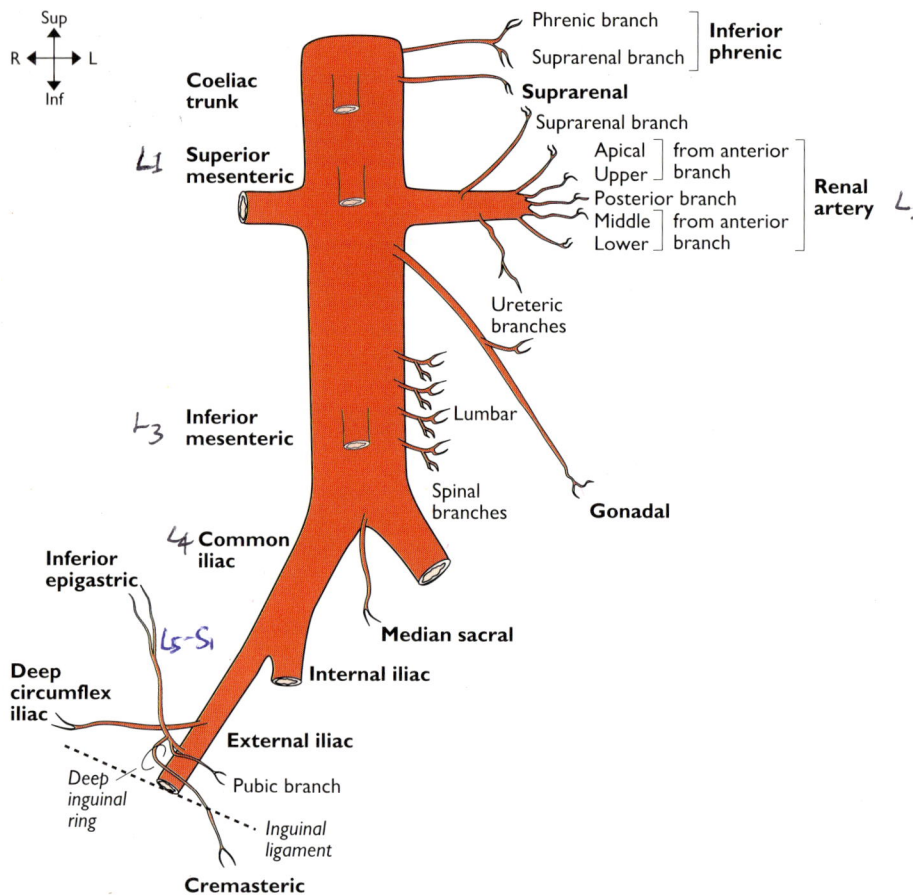

Abdominal aorta & external iliac artery

ABDOMINAL AORTA
From: Thoracic aorta
To: Common iliac arts

This main artery arises as the continuation of the thoracic aorta as it passes, in the midline, posterior to the median arcuate ligament of the diaphragm at T12 and it ends slightly to the left of the midline at L4 where it terminates as the left and right common iliac arteries. Anterior to it, from above downwards, are the coeliac trunk and its branches, the coeliac plexus, lesser sac, superior mesenteric artery, left renal vein, body of pancreas, commencement of each gonadal artery, fourth part of the duodenum, posterior parietal peritoneum, attachment of the mesentery and inferior mesenteric artery. Posterior to it are the lumbar arteries and left lumbar veins, anterior longitudinal ligament and vertebral bodies with their intervertebral discs. To its right are the cisterna chyli, thoracic duct, azygos vein, right crus of diaphragm and inferior vena cava. To its left are the left crus of diaphragm, left coeliac ganglion, the duodenojejunal flexure (upper border of L2), sympathetic trunk and inferior mesenteric vessels. On both sides the phrenic, suprarenal and renal vessels are lateral whilst distal to the bifurcation of the abdominal aorta is the median sacral artery.

Common iliac arteries. These commence at the bifurcation of the abdominal aorta just to the left of the midline at L4 and pass inferolaterally to the level between L5 and S1 vertebrae where they bifurcate anterior to the sacro-iliac joint to give the external and internal iliac arteries. Anterior to each vessel are sympathetic contributions to the superior hypogastric plexus, the ureter (near the terminal bifurcation of the vessel), peritoneum and small bowel. In addition on the left side the superior rectal artery lies anterior. Posterior to each vessel are the sympathetic trunk, obturator nerve, lumbosacral trunk, iliolumbar artery and the bodies of L4 and L5 with the disc between them. In addition posteriorly on the right side are the terminal portions of the common iliac veins and the commencement of the inferior vena cava. The left common iliac vein lies posteromedial to the left common iliac artery. Psoas major lies lateral to each vessel.

Gonadal artery. Descends passing obliquely inferiorly on the posterior abdominal wall to the level of the external iliac artery. The testicular arteries pass around the lower border of the false pelvis to enter the inguinal canal through the deep ring to form part of the spermatic cord. The ovarian vessels descend over the external iliac vessels into the infundibulopelvic fold to supply the ovary via the broad ligament. The common relations of the arteries in both sexes are: left, posterior — psoas, genitofemoral nerve, ureter and external iliac artery. Left, anterior — inferior mesenteric vein, left colic artery and sigmoid mesentery. Right, posterior — inferior vena cava, psoas, genitofemoral nerve, ureter and external iliac artery. Right, anterior — third part of the duodenum, right colic artery and ileal mesentery.

EXTERNAL ILIAC ARTERY
From: Common iliac art
To: Femoral art

The external iliac artery descends laterally from the common iliac artery to pass under the inguinal ligament at the mid inguinal point (half way between the anterior superior iliac spine and symphysis pubis) where it becomes the femoral artery.

Posterior and lateral to it is the medial border of psoas major whilst the femoral vein comes to lie medially. Anteromedially it is covered by peritoneum on which lies small bowel with sigmoid colon additionally on the left. It is crossed at its origin by the ureter and then by the gonadal vessels, genital branch of the genitofemoral nerve, deep circumflex iliac vein and vas deferens or round ligament.

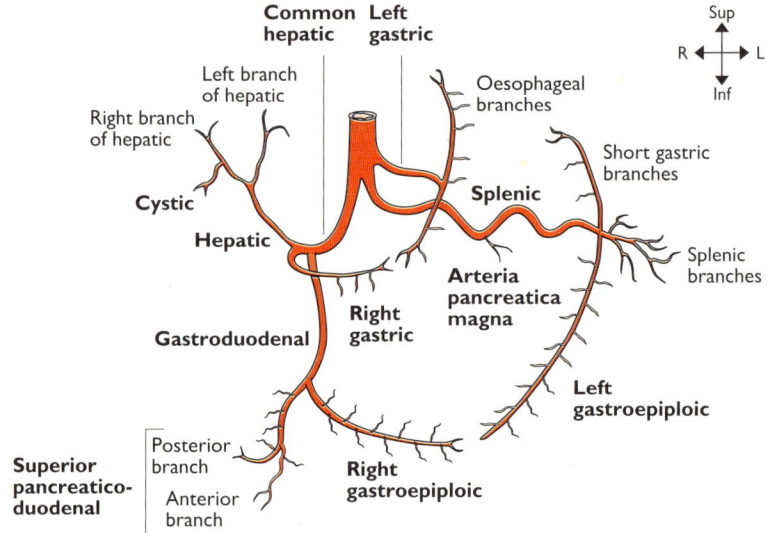

Coeliac trunk

COELIAC TRUNK

**From: Abdominal aorta at lower border
of T12**

**To: Terminal brs — left gastric, splenic
& common hepatic arts**

The coeliac trunk (axis) arises from the anterior aspect of the abdominal aorta at the level of the lower border of T12 and after 1 cm divides into its three terminal branches.

Left gastric artery. Passes superolaterally on the posterior wall of the lesser sac to reach the apex of this structure at the cardio-oesophageal junction where it divides into oesophageal branches to supply the lower third of the oesophagus through the oesophageal opening in the diaphragm. Its terminal gastric branches run inferiorly along the upper portion of the lesser curve of the stomach to anastomose with the right gastric artery.

Splenic artery. Passes laterally to the left, angled slightly superiorly, running in the posterior wall of the lesser sac. Its course is markedly tortuous as it runs along the superior border of the pancreas, passing anterior to the left crus of the diaphragm, the upper pole of the left kidney and the left suprarenal gland before entering the lienorenal ligament to reach to the splenic hilum.

Common hepatic artery. Runs infero-laterally to the right in the posterior wall of the lesser sac towards the first part of the duodenum where it gives off the gastroduodenal and right gastric arteries. It then curves anteriorly as the hepatic artery to pass into the peritoneal reflection which forms the inferior margin of the opening of the lesser sac. It approaches the portal vein from its left side and then comes to lie anterior to it as it ascends in the free border of the lesser omentum (anterior margin of the foramen of Winslow or epiploic foramen) before terminating at the porta hepatis as the right and left hepatic branches.

Gastroduodenal artery. Descends directly behind the first part of the duodenum to the left of the common bile duct and divides at the upper border of the pancreas into terminal branches.

Right gastric artery. Arises from the hepatic artery as it enters the lesser omentum and passes along the lesser curve of the stomach to anastomose with the left gastric artery.

I

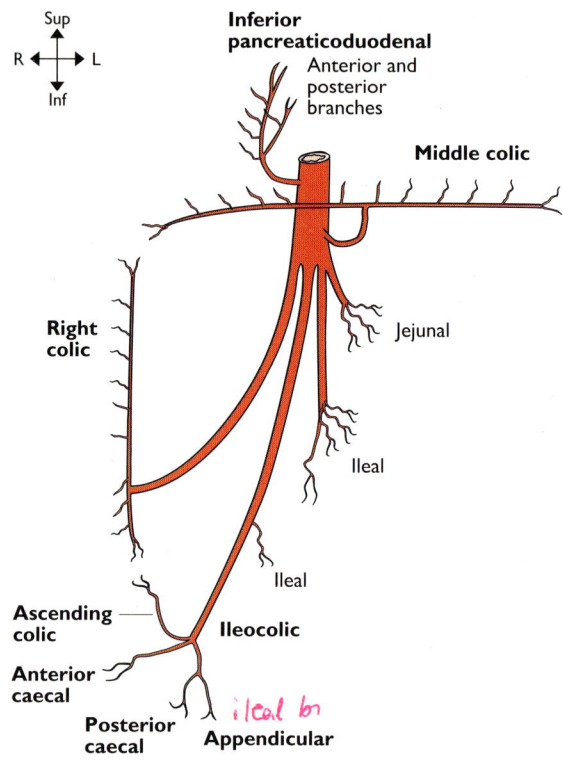

Superior mesenteric artery

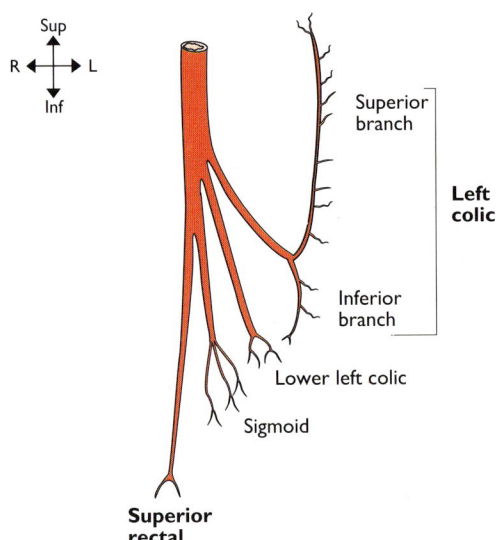

Inferior mesenteric artery

SUPERIOR MESENTERIC ARTERY
From: Abdominal aorta
To: Terminal brs

The superior mesenteric artery arises from the anterior surface of the abdominal aorta at the level of L1. It passes inferiorly over the left renal vein with the splenic vein and body of pancreas anterior to it. It next lies on the uncinate process of the pancreas and the junction of the third and fourth parts of the duodenum from where it passes obliquely and to the right into the mesentery of the small bowel before giving off its terminal branches. The superior mesenteric vein is on its right whilst posterior to the terminal branches are the inferior vena cava, the right ureter and psoas major. The superior mesenteric plexus of nerves surrounds the artery. It supplies bowel from the mid second part of the duodenum, jejunum, ileum, ascending and right two thirds of transverse colon.

Inferior pancreaticoduodenal artery. Leaves the superior mesenteric artery as it begins to cross the duodenum and divides into an anterior and posterior branch. The anterior branch passes to the right to anastomose with the anterior superior pancreaticoduodenal artery anterior to the head of the pancreas. The posterior branch passes also to the right but posterior to the head of the pancreas to anastomose with the posterior superior pancreaticoduodenal artery.

Ileocolic artery. Passes obliquely inferiorly to the right in the root of the mesentery where it passes anterior to the right ureter and right gonadal vessels to reach the caecum where it divides into its terminal branches.

INFERIOR MESENTERIC ARTERY
From: Abdominal aorta
To: Terminal brs

This artery arises from the anterior surface of the abdominal aorta at the level of L3 posterior to the third and fourth part of the duodenum. It passes inferiorly and to the left crossing the left common iliac artery medial to the left ureter. The inferior mesenteric vein lies on its left (lateral) side. It divides into its terminal branches in the descending mesocolon. It supplies the left third of the transverse colon, the descending and sigmoid colon and the rectum to the dentate line of the anus.

I

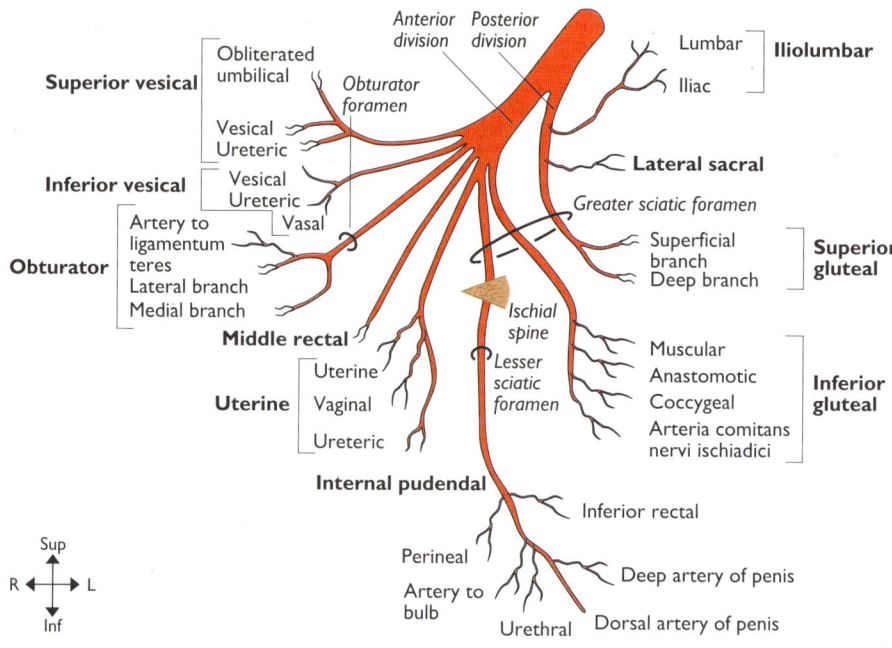

Internal iliac artery

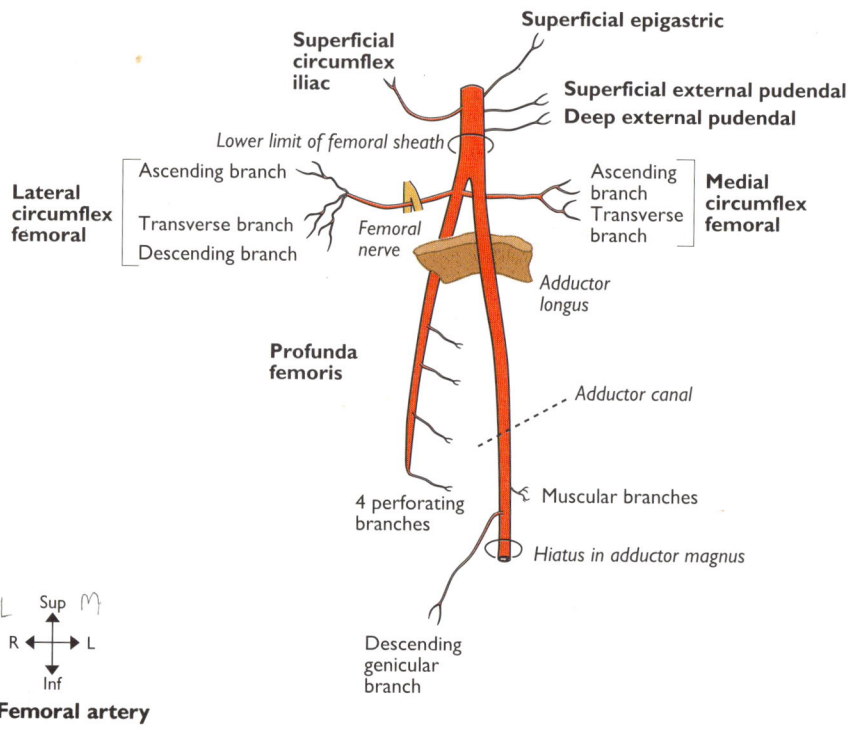

Femoral artery

INTERNAL ILIAC ARTERY
From: **Common iliac art**
To: **Terminal brs**

The artery commences at the level of the disc between L5 and S1 and passes posteriorly into the pelvis for 4 cm before forming anterior and posterior divisions which break up into their terminal branches. Anterior lie the ureter and fallopian tube and ovary in the female. Posterior are the internal iliac vein, lumbosacral trunk and sacro-iliac joint. On the lateral side are the external iliac artery and vein, obturator nerve and psoas major. The parietal peritoneum and small bowel lie medially.

Internal pudendal artery. Arises from the anterior division of the internal iliac artery and descends on the lateral wall of the pelvis towards the greater sciatic foramen. It leaves the pelvis via this foramen, inferior to piriformis, before passing over the tip of the ischial spine to enter the ischio-anal fossa via the lesser sciatic foramen. In runs on the lateral wall of the ischio-anal fossa on obturator internus in the pudendal (Alcock's) canal. It passes into the deep perineal pouch where it gives off its terminal branches. (Other branches of the perineal branch, not illustrated, are transverse perineal and posterior scrotal.)

Note: In the female the vaginal artery is equivalent to the inferior vesical artery in the male, and the uterine artery is equivalent to the middle rectal artery. The round ligament is supplied by the uterine artery whilst the vas deferens is usually supplied by the inferior vesical or less often by the superior vesical artery.

FEMORAL ARTERY
From: **External iliac art**
To: **Popliteal art**

This is the continuation of the external iliac artery and commences posterior to the inguinal ligament at the mid-inguinal point (half way between the anterior superior iliac spine and the symphysis pubis). It ends as it passes through the adductor hiatus in adductor magnus to become the popliteal artery. It emerges from under the inguinal ligament with the femoral vein medial to it, both within the femoral sheath. Lateral to it and outside the femoral sheath is the femoral nerve. It lies on the tendon of psoas major and is separated from pectineus and adductor longus by the femoral vein which comes to lie progressively more posterior to the artery within the femoral triangle. As the femoral artery enters the adductor canal it lies on adductor longus then adductor magnus. It is covered initially only by deep fascia then by sartorius; the saphenous nerve passes anteriorly from lateral to medial. Anterolateral to the artery is vastus medialis.

Profunda femoris is the main branch of the femoral artery which is given off postero-laterally just below the femoral sheath 3.5 cm below the inguinal ligament. It runs posteriorly between pectineus and adductor longus to pass into the deep thigh where it provides the deep structures and the posterior and medial compartments with their main arterial supply. Perforating and descending branches anastomose with the genicular branches of the popliteal artery.

I

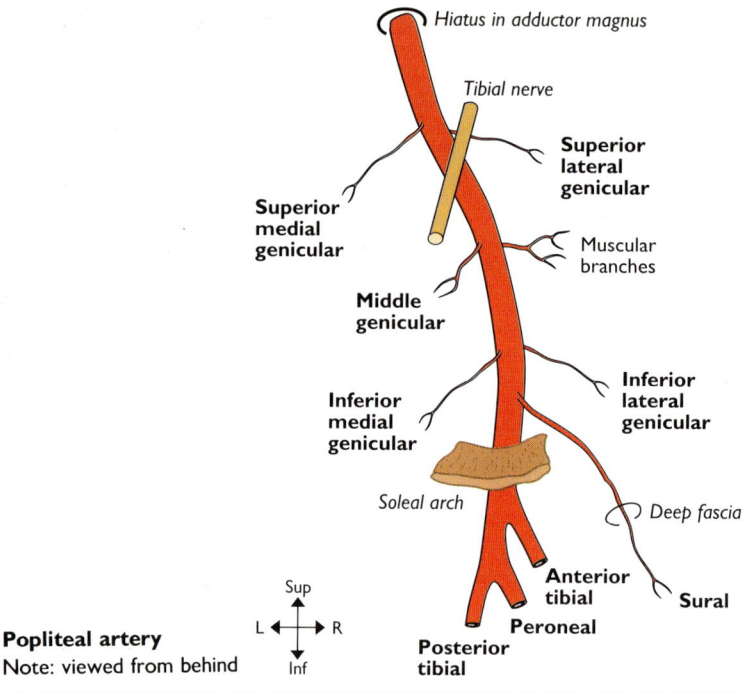

Hiatus in adductor magnus

Tibial nerve

Superior lateral genicular

Superior medial genicular

Muscular branches

Middle genicular

Inferior medial genicular

Inferior lateral genicular

Soleal arch

Deep fascia

Anterior tibial

Sural

Peroneal

Posterior tibial

Popliteal artery
Note: viewed from behind

Sup
L ← → R
Inf

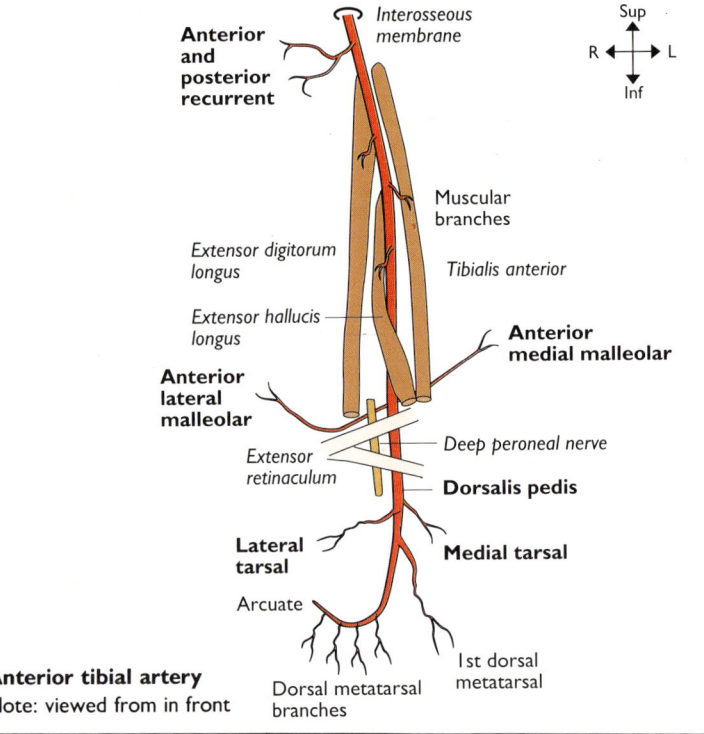

Interosseous membrane

Anterior and posterior recurrent

Sup
R ← → L
Inf

Muscular branches

Extensor digitorum longus

Tibialis anterior

Extensor hallucis longus

Anterior medial malleolar

Anterior lateral malleolar

Deep peroneal nerve

Extensor retinaculum

Dorsalis pedis

Lateral tarsal

Medial tarsal

Arcuate

1st dorsal metatarsal

Anterior tibial artery
Note: viewed from in front

Dorsal metatarsal branches

I

POPLITEAL ARTERY
From: Femoral art
To: Ant & post tibial arts

This artery commences as the continuation of the femoral artery as the latter passes through the hiatus in adductor magnus and ends as it passes under the fibrous arch of soleus where it immediately divides into anterior and posterior tibial arteries. The popliteal artery extends from a hand's breadth above the knee and to the same distance below it. It enters the popliteal fossa medial to the femur and becomes the deepest structure, lying with only fat between it and the popliteal surface of the femur. Lower down it lies on the capsule of the knee joint and then on popliteus. Biceps femoris is lateral to it and semimembranosus medial. Lower down it lies between the two heads of gastrocnemius. It is crossed laterally to medially by the tibial nerve and the popliteal vein with the vein always between the artery and nerve.

ANTERIOR TIBIAL ARTERY
From: Popliteal art
To: Dorsalis pedis art

This artery commences at the bifurcation of the popliteal artery just under the fibrous arch of soleus, at the distal border of popliteus. It supplies the structures in the extensor compartment of the lower leg. It passes anteriorly between the heads of tibialis posterior to pass above the upper border of the interosseous membrane, medial to the neck of the fibula accompanied by its venae commitantes. It descends on the interosseous membrane and crosses the lower tibia at the ankle joint, mid way between the malleoli and there becomes the dorsalis pedis artery. Initially it lies between tibialis anterior (medially) and extensor digitorum longus (laterally), then between tibialis anterior and extensor hallucis longus. At the ankle it is crossed anteriorly by the extensor retinacula and also from lateral to medial by the tendon of extensor hallucis longus. The deep peroneal nerve is initially lateral to the artery but passes anterior to it before again becoming lateral. The anterior tibial veins run in close association with the artery throughout.

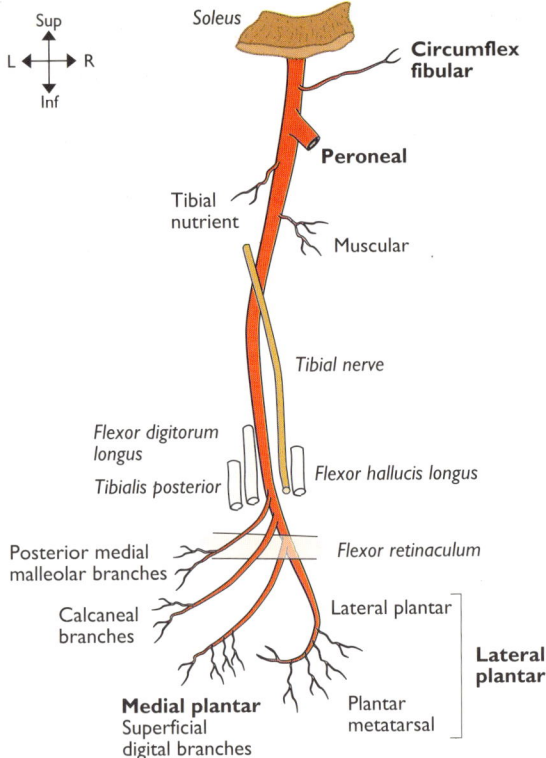

Posterior tibial artery
Note: viewed from behind

POSTERIOR TIBIAL ARTERY
From: Popliteal art
To: Med & lat plantar arts

This artery arises at the bifurcation of the popliteal artery just under the fibrous arch of soleus, at the lower border of popliteus, and ends by bifurcating into the medial and lateral plantar arteries deep to abductor hallucis. It supplies structures in the posterior compartment of the lower leg. It is accompanied by venae commitantes and lies from above downwards on tibialis posterior, flexor digitorum longus, the tibia and the ankle joint. It lies deep to gastrocnemius, soleus, the flexor retinaculum and abductor hallucis. Posterior to the medial malleolus it lies between the tendon of flexor digitorum longus and the tibial nerve which crosses posterior to the artery mid way down the calf from the medial side to become postero-lateral. (Other branch, not illustrated, is a communicating branch to peroneal artery.)

I

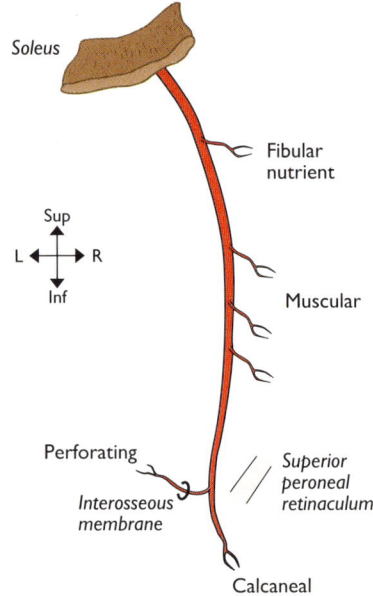

Soleus

Fibular
nutrient

Sup

L ← → R

Inf

Muscular

Perforating

Interosseous
membrane

Superior
peroneal
retinaculum

Calcaneal

Peroneal artery

PERONEAL ARTERY
From: Post tibial art
To: Terminal brs

This is a branch of the posterior tibial artery arising 2.5 cm below its origin under soleus. It supplies structures in the lateral compartment of the lower leg. Due to the proximity of origin of the three terminal branches of the popliteal artery this point of origin is commonly referred to as the 'popliteal trifurcation'. It passes infero-laterally to reach and run along the medial crest of the fibula between tibialis posterior and flexor hallucis longus to divide into its terminal branches at the level of the inferior tibiofibular joint and the superior peroneal retinaculum. Above, it is covered by soleus and deep fascia whilst in the lower leg flexor hallucis longus crosses it from lateral to medial.

2: VEINS

2

2

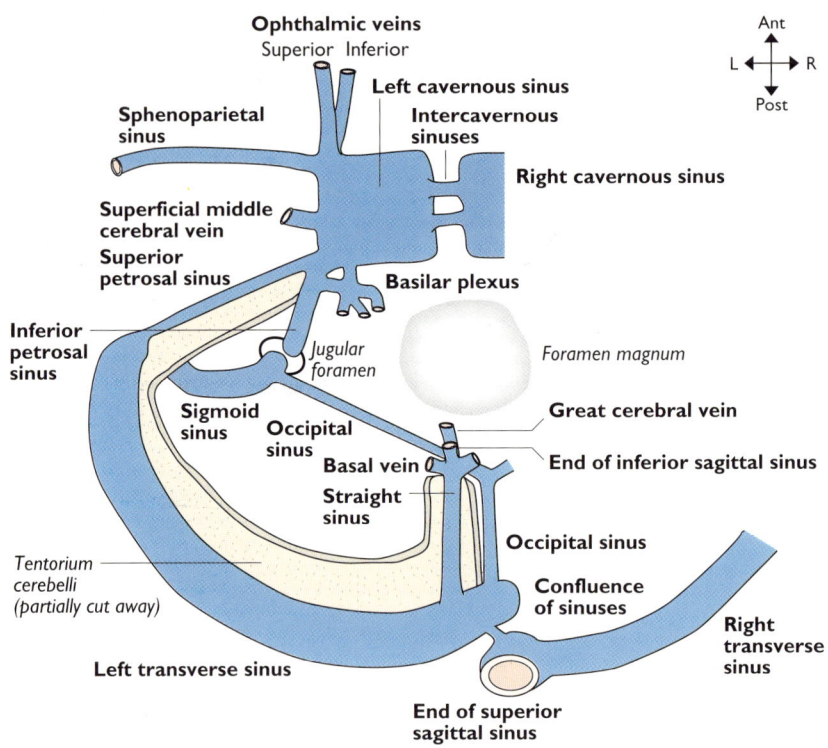

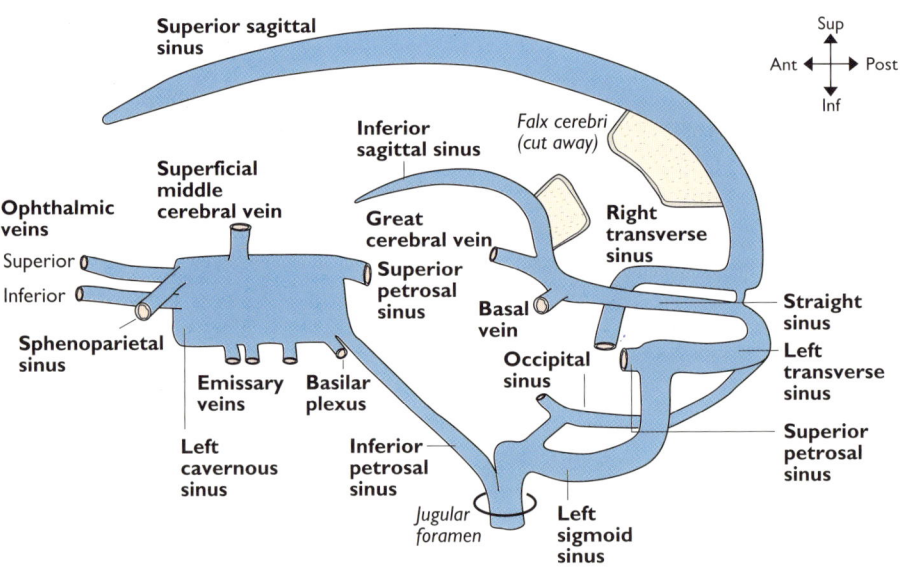

Intracranial sinuses & veins

INTRACRANIAL SINUSES & VEINS

From: Cerebrum, cerebellum & diploe of the skull

To: Internal jugular V

The cerebrum, cerebellum and bones of the skull are drained by the external, internal and meningeal veins to the sinuses. The sinuses lie within the inner layer of the dura mater, either as an endothelial lined space in its free edge (inferior sagittal and straight sinuses), or a similarly lined space where the dura is reflected over the bone of the inner surface of the skull. They are characteristically thin walled, contain no valves and communicate freely with each other.

Superior sagittal sinus. Lies in the superior margin of the falx cerebri draining the arachnoid granulations as it does so and usually drains as a continuation into the right transverse sinus. It frequently connects at its termination with the left transverse sinus.

Inferior sagittal sinus. Runs in the inferior free margin of the falx cerebri draining medial cortical veins as it does so, and terminates by fusing with the great cerebral vein of Galen and right and left basal veins to form the straight sinus.

Straight sinus. Runs in the junction of the falx cerebri and tentorium cerebelli for a short distance before terminating in its continuation — the left transverse sinus.

Transverse (lateral) sinus. Runs in the lateral border of the tentorium cerebelli grooving the occipital and squamous temporal bones, to terminate in the sigmoid sinus as it receives the superior petrosal sinus from the cavernous sinus on each side.

Sigmoid sinus. Deeply grooves the temporal bone as it passes inferomedially into the posterior compartment of the jugular foramen at the inferior margin of which it

unites with the inferior petrosal sinus to form the internal jugular vein.

Cavernous sinus. Lies on the lateral wall of the body of the sphenoid bone and is a lateral relation of the sella turcica, the pituitary gland and the sphenoidal air sinus. It lies medial to the medial gyrus of the temporal lobe and lies on the greater wing of the sphenoid bone. Lying in it is the internal carotid artery (carotid syphon) with the abducent nerve (VI) on its lateral surface and lying on its lateral wall are nerves (from above down): oculomotor (III), trochlear (IV), ophthalmic (Va) and maxillary (Vb) divisions of the trigeminal. It has a sponge-like reticular structure and its connections, particularly those with the other major sinuses (as shown opposite), frequently provide both supply to, and drainage from, the sinus. There are two intercavernous sinuses connecting the cavernous sinuses to each other.

Occipital sinus. Begins at the foramen magnum and ascends to end in the confluence of sinuses.

Confluence of sinuses is at the lowest, posterior end of the superior sagittal sinus at the point that it turns, usually to the right, to become the transverse sinus. It connects with the straight, occipital and opposite transverse sinuses.

Sphenoparietal sinus. Runs along the lesser wing of the sphenoid bone and drains into the cavernous sinus.

Superior petrosal sinus. Runs along the petrous temporal bone where the edge of the tentorium cerebelli attaches and hence connects the cavernous and transverse sinuses.

Inferior petrosal sinus. Runs inferiorly to connect the cavernous sinus to the internal jugular vein.

2

2

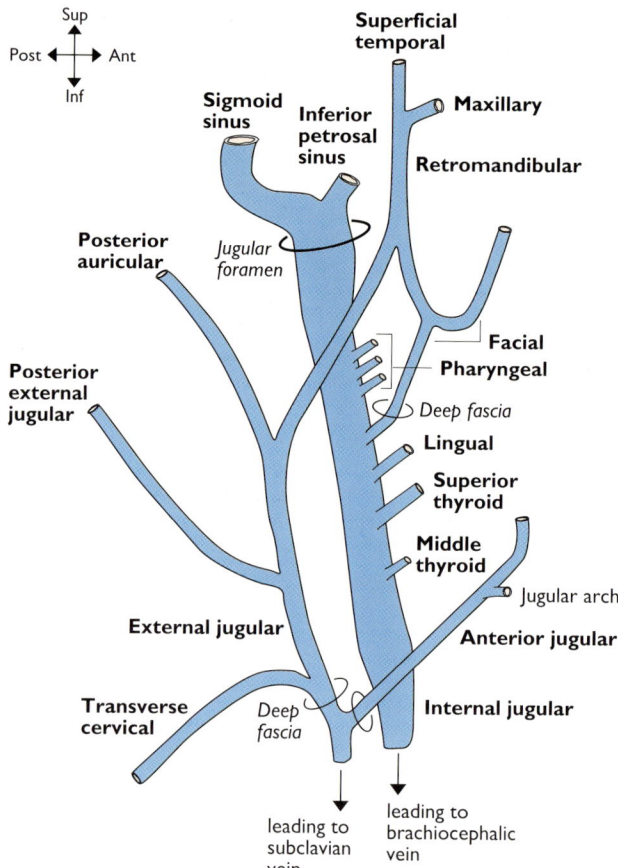

Sup

Post ◀━━▶ Ant

Inf

Superficial temporal

Sigmoid sinus

Inferior petrosal sinus

Maxillary

Retromandibular

Posterior auricular

Jugular foramen

Facial

Pharyngeal

Posterior external jugular

Deep fascia

Lingual

Superior thyroid

Middle thyroid

Jugular arch

External jugular

Transverse cervical

Deep fascia

Anterior jugular

Internal jugular

leading to subclavian vein

leading to brachiocephalic vein

Internal & external jugular veins

INTERNAL JUGULAR VEIN
From: **Sigmoid & inf petrosal sinuses**
To: **Brachiocephalic Vs**

It runs almost vertically downwards within the carotid sheath although its covering is thin and readily stretched. Its relationship to the internal carotid artery is as follows: posterior to the artery at the level of C2, posterolateral at C3 and lateral at C4, the vagus nerve (X) lies between the two throughout. Outside the sheath it is surrounded by deep cervical lymph nodes and it lies on (from above down): the lateral mass of the atlas (C1), prevertebral fascia, scalenus medius, scalenus anterior and the dome of the cervical pleura. It is crossed at its origin by the spinal accessory nerve, the upper root of the ansa cervicalis in its middle third and is overlaid by sternocleidomastoid and the tendon of omohyoid in its lower third.

EXTERNAL JUGULAR VEIN
From: **Various brs**
To: **Subclavian V**

It arises from tributaries as shown opposite and drains into the subclavian vein. The external jugular system lies within the superficial tissues of the neck (as does the anterior jugular system). The external and anterior jugular veins pierce the deep fascia of the neck, usually posterior to the clavicular head of sternocleidomastoid to fuse before draining into the subclavian vein.

2

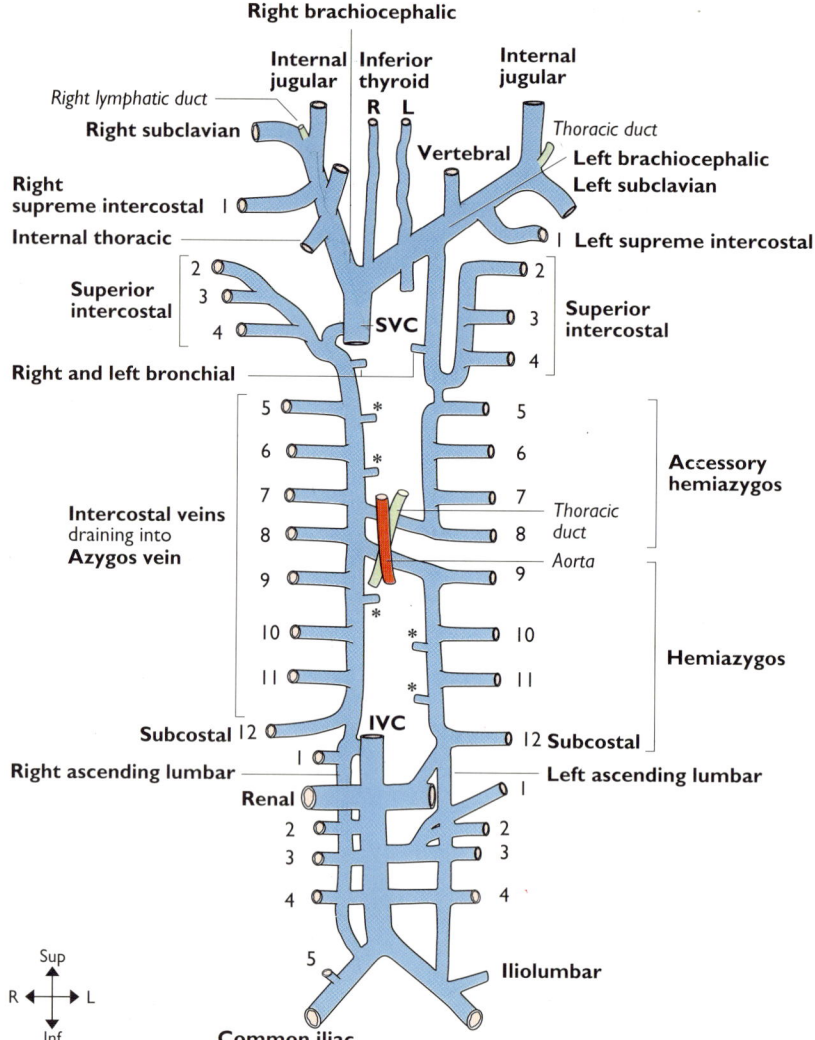

Superior vena cava & azygos veins

Note: (1) Accessory hemiazygos crosses to azygos at T7 and hemiazygos at T8 — each crosses behind thoracic aorta, oesophagus and thoracic duct, (2) left bronchial vein may enter accessory hemiazygos. * = oesophageal and mediastinal veins, IVC = inferior vena cava, SVC = superior vena cava

SUPERIOR VENA CAVA
From: Brachiocephalic Vs
To: Right atrium

It is formed posterior to the right first costal cartilage and passes posterior to the right sternal border where it is a close posterior relation of the right internal thoracic vessels and sternal periosteum and is occasionally overlaid by the anterior segment of the right upper lobe of the lung. It lies anterolateral to the trachea and upper right lung hilum with the right phrenic nerve applied to its right lateral surface. It receives the azygos vein into its posterior surface at the level of T4. It enters the superior surface of the right atrium without any valvular mechanism guarding its orifice.

Left brachiocephalic vein. Formed posterior to the left sternoclavicular joint and anterior to the cervical pleura. It passes obliquely downwards to the right, posterior to the manubrium, separated from it only by the thymus gland or its remnant. It lies anterior to the left common carotid and brachio-cephalic arteries and crosses the upper anterior aortic arch. Other tributaries are thymic and pericardial veins.

Right brachiocephalic vein. Formed posterior to the right sternoclavicular joint and passes directly inferiorly behind the right side of the manubrium sterni, anterolateral to the trachea and antero-medial to the pleura over the upper lobe of the lung.

AZYGOS VEINS
From: Inf vena cava/ascending lumbar Vs
To: Sup vena cava

The azygos veins drain the upper lumbar region and the thoracic wall. There is a single system on the right whilst on the left

there are two — the hemiazygos and accessory hemiazygos that drain over into the azygos separately.

Azygos vein. Arises at the approximate level of the right renal vein either as a posterior tributary of the inferior vena cava or as a confluence of the right ascending lumbar and right subcostal vein. It passes through the aortic opening of the diaphragm under the right crus at the level of T12 vertebra and ascends on the right side of the vertebral bodies behind the oesophagus. It turns anteriorly to pass over the hilum of the right lung, lateral to the oesophagus, trachea and right vagus, to enter the superior vena cava at the level of T4. Its tributaries are the lower eight right posterior intercostal veins, the right superior intercostal vein (draining the 2nd, 3rd and 4th right intercostal veins), bronchial and oesophageal veins and, from the left side, the two hemiazygos veins.

Hemiazygos vein. Arises from the confluence of the left ascending lumbar vein, the left subcostal vein and often a tributary from the left renal vein. It ascends through the aortic opening of the diaphragm and onto the left side of the thoracic vertebra to the level of T9 from where it crosses posterior to the aorta, oesophagus and thoracic duct to enter the azygos vein at T8. It drains the four lower left posterior intercostal veins (9−12th).

Accessory hemiazygos vein. Drains the 5−8th left posterior intercostal veins and runs inferiorly on the left side of the vertebral bodies to T8 where it crosses similarly to the hemiazygos vein to enter the azygos vein at T7. It also receives tributaries from the bronchial and mid-oesophageal veins.

Note: The anterior intercostal veins drain to the musculophrenic and internal thoracic veins.

2

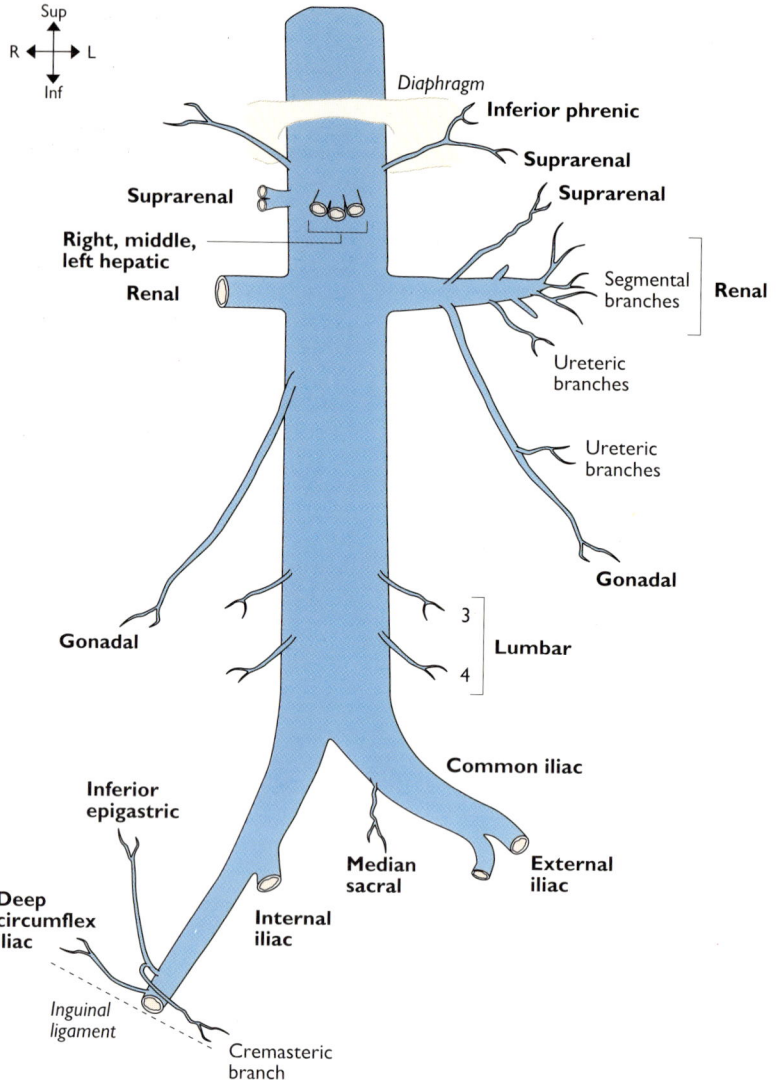

Sup

R ←→ L

Inf

Diaphragm

Inferior phrenic

Suprarenal

Suprarenal

Suprarenal

Right, middle, left hepatic

Renal

Segmental branches

Renal

Ureteric branches

Ureteric branches

Gonadal

3

4

Lumbar

Gonadal

Common iliac

Inferior epigastric

Median sacral

External iliac

Deep circumflex iliac

Internal iliac

Inguinal ligament

Cremasteric branch

Inferior vena cava

INFERIOR VENA CAVA
From: **Common iliac Vs**
To: **Right atrium**

It arises as the fusion of the common iliac veins anterolateral to the L5 vertebral body lying posterior to the right common iliac artery. It ascends, initially posterolateral to, then lateral to the aorta and lies anterolateral to the right of the bodies of L5–L1. It lies on (from below up): right lumbar arteries, right renal artery, right sympathetic chain, right suprarenal gland, right crus of the diaphragm and right phrenic artery. It is crossed by (from below up): the root of the ileal mesentry, the third part of the duodenum, the head of the pancreas, the common bile duct, the portal vein, the first part of the duodenum, the posterior abdominal peritoneum and the bare area of the liver. It is hugged closely on its right side by the right suprarenal gland and forms the posterior wall of the epiploic foramen of Winslow below this. After passing through the caval orifice in the diaphragm (T8) with the right phrenic nerve lateral to it, it runs for a short distance within the middle mediastinum before entering the inferior aspect of the right atrium. The vein possesses a 'valve-like' flap guarding the medial portion of its orifice.

Lumbar veins. Drain somewhat inconsistently but usually the 3rd and 4th drain directly into the inferior vena cava whilst above this level they drain into the ascending lumbar veins and hence to the azygos and hemiazygos systems. There are, however, usually connections of the 3rd and 4th lumbar veins with the ascending lumbar veins.

2

2

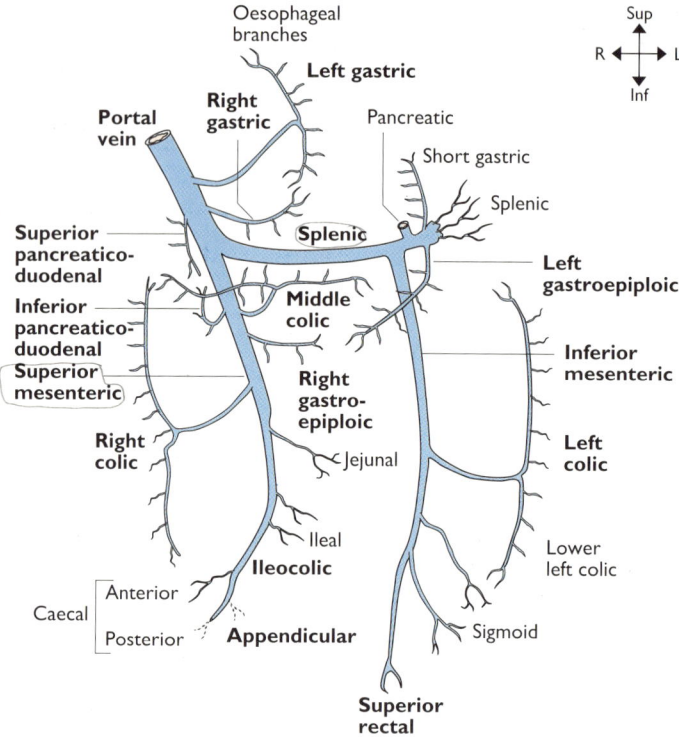

Oesophageal
branches

Left gastric

**Right
gastric**

**Portal
vein**

Pancreatic

Short gastric

Splenic

Sup

R ← → L

Inf

**Superior
pancreatico-
duodenal**

Splenic

**Left
gastroepiploic**

**Inferior
pancreatico-
duodenal**

**Middle
colic**

**Superior
mesenteric**

**Right
gastro-
epiploic**

**Inferior
mesenteric**

**Right
colic**

Jejunal

**Left
colic**

Ileal

Lower
left colic

Caecal

Anterior

Ileocolic

Posterior

Appendicular

Sigmoid

**Superior
rectal**

Portal vein

PORTAL VEIN
From: **Sup mesenteric & splenic Vs**
To: **Porta hepatis**

The vein is formed from the union of the superior mesenteric and splenic veins at the level of L2 just at the right of the midline. At its formation it lies posterior to the neck of the pancreas and anterior to the inferior vena cava. It runs superiorly inclining slightly to the right lying posterior to the first part of the duodenum and anterior to the inferior vena cava. The common bile duct comes to lie anterolateral to it from the right and the hepatic artery comes to lie anteromedial to it from the left. Ascending in the free border of the lesser omentum it continues to lie posterior to these two structures and forms the anterior margin of the aditus to the lesser sac (foramen of Winslow). It divides into terminal right and left branches as it enters the porta hepatis.

Portosystemic anastomoses
1 Lower end of oesophagus. The veins from the lower third of the oesophagus drain downwards to the left gastric vein (portal) and, above this level, oesophageal veins drain to the azygos and hemiazygos systems (systemic).

2 Upper end of anal canal. At the anal columns in the upper half of the anal canal there is a venous watershed between the drainage above by the superior rectal veins (portal via inferior mesenteric vein) and drainage below by the inferior and middle rectal veins (systemic via the pudendal and internal iliac veins).

3 Bare area of liver. Where the bare area of the liver lies in contact with the diaphragm there is a watershed of venous drainage between the hepatic veins (portal) and the phrenic veins (systemic).

4 Periumbilical. The ligamentum teres represents a venous watershed in that its upper portion is drained into the portal system (via the left branch of the portal vein) whilst its lower portion is drained into the systemic system (indirectly via the great saphenous and axillary veins).

5 Retroperitoneal. Branches of the left and right colic and splenic veins (portal) may meet branches of the lumbar veins (systemic via inferior vena cava and azygos systems) in the retroperitoneal area.

2

2

3: LYMPHATICS

Note: Deep lymphatics follow arteries and superficial ones follow veins.

3

3

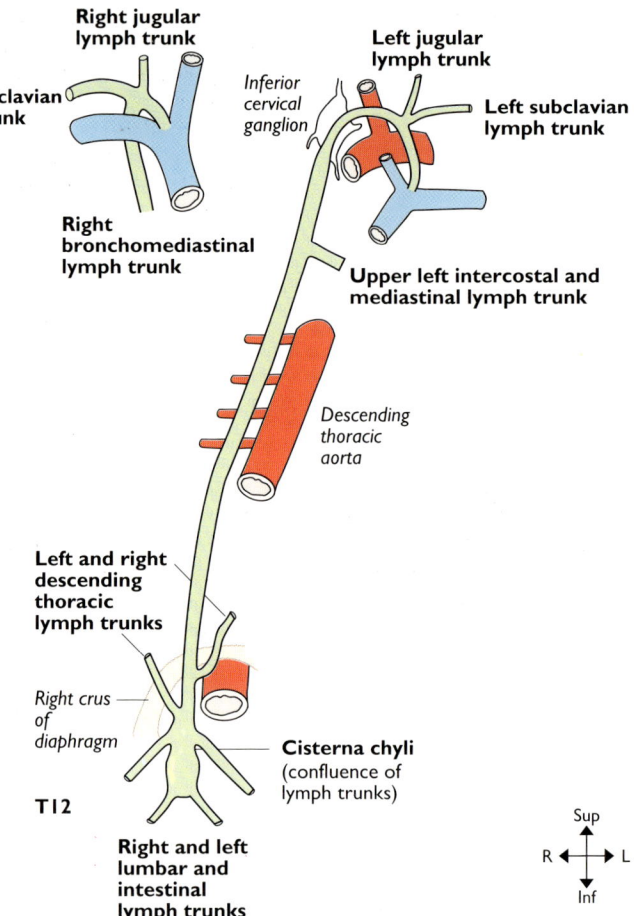

**Right jugular
lymph trunk**

**Right subclavian
lymph trunk**

*Inferior
cervical
ganglion*

**Left jugular
lymph trunk**

**Left subclavian
lymph trunk**

**Right
bronchomediastinal
lymph trunk**

**Upper left intercostal and
mediastinal lymph trunk**

*Descending
thoracic
aorta*

**Left and right
descending
thoracic
lymph trunks**

*Right crus
of
diaphragm*

T12

Cisterna chyli
(confluence of
lymph trunks)

**Right and left
lumbar and
intestinal
lymph trunks**

Sup

R ← → L

Inf

Thoracic & right lymphatic ducts

THORACIC & RIGHT LYMPHATIC DUCTS
From: **Cisterna chyli**
To: **Left subclavian V**

Thoracic duct
Receives:
- Left jugular trunk
- Left subclavian trunk
- Cisterna chyli
- Most thoracic lymphatics

Drains:
- All body tissue below the diaphragm
- Left arm
- Left head and neck
- Left thorax
- Lower right thorax

It originates from the upper cisterna chyli on the right anterolateral side of to the body of T12, lying lateral to the abdominal aorta. It passes posterior to the right crus of the diaphragm and ascends on the right posterior intercostal arteries with the aorta on its left and the azygos vein on its right. It slopes to the left in the mid thorax crossing the vertebral column posterior to the oesophagus at the level of T5. It continues superiorly to the left of the vertebral column, posterolateral to the oesophagus and then posterior to the dome of the cervical pleura as it passes anterior to the inferior cervical (stellate) ganglion. It arches anteriorly over the vertebral and subclavian arteries and over the pleura to reach the posterosuperior aspect of the left subclavian vein as the latter joins the left internal jugular vein.

Right lymphatic trunk
Receives:
- Right subclavian trunk
- Right jugular trunk
- Right bronchomediastinal trunk

Drains:
- Right head and neck
- Right arm
- Right upper thorax

The three trunks usually drain separately but the first two often join to give a right lymphatic duct which ends in the right subclavian vein. In which case it has a very short course from its formation anterior to scalenus anterior, passing over the dome of the cervical pleura to reach the right subclavian vein as the latter joins with the right internal jugular vein.

3

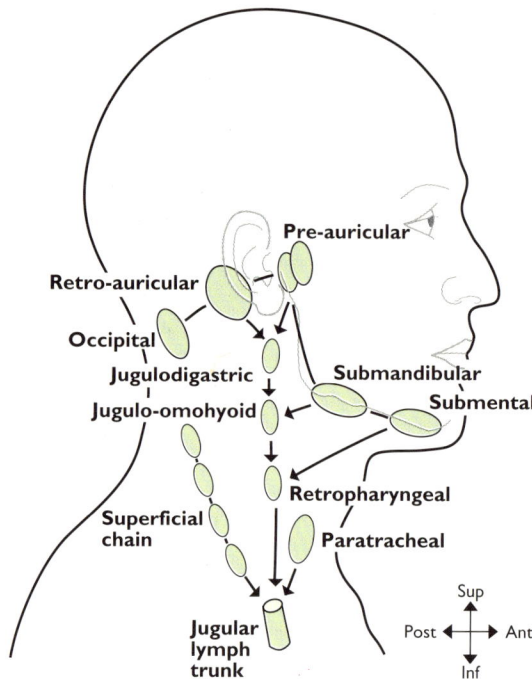

Lymph glands — head & neck

LYMPH GLANDS — HEAD & NECK

Circular nodal chain
Submental (bilateral)
  Anterior tongue
  Floor of mouth
  Lower incisor and canine teeth
  Lower lip
  Skin of anterior chin
Submandibular
  Upper lip
  Cheek
  Nose
  Forehead and anterior scalp
  Middle tongue
  Lower molar and premolar teeth
  All upper teeth
  Sublingual gland
  Submandibular gland
  Anterior half of nasal cavity and nasal
    sinuses
Deep and superficial pre-auricular (parotid)
  Middle scalp
  Skin of temple
  Pinna
  Parotid gland
  Posterior orbit

Retro-auricular (mastoid)
  Pinna
  Posterior scalp
Occipital
  Posterior scalp

Deep cervical chain
Jugulodigastric
  Palatine tonsil
  Upper pharynx
  Posterior tongue
Jugulo-omohyoid
  Posterior half of nasal cavity and nasal
    sinuses
  Palate (hard and soft)
Retropharyngeal
  Pharynx
Paratracheal
  Hypopharynx
  Larynx
  Trachea
  Thyroid
  Parathyroids

Superficial cervical chain
Skin of neck

3

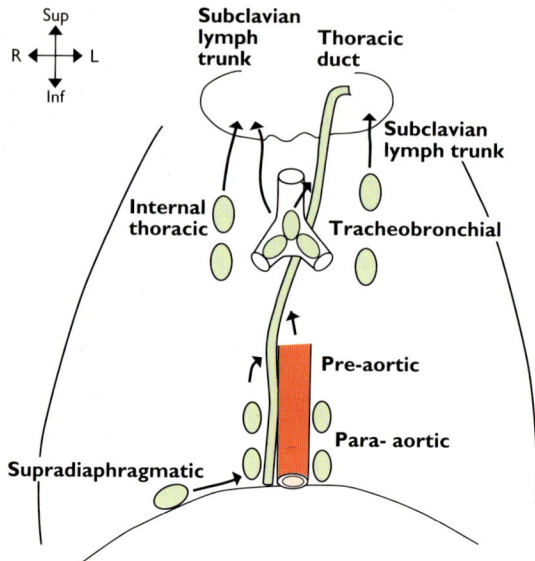

Lymph glands — thorax

LYMPH GLANDS – THORAX

Pre-aortic
 Middle third of oesophagus
Supradiaphragmatic
 Diaphragm
 Subphrenic spaces
 Bare area of liver
Tracheobronchial
 Heart and all layers of pericardium
 Lungs and visceral pleura
 Extrapulmonary bronchi
 Trachea
 Thymus (occasionally thyroid isthmus)
Para-aortic
 Thoracic wall
 Parietal pleura
 Anterior abdominal wall
Internal thoracic
 Breast
 Anterior thoracic wall
 Upper abdominal muscles
 Diaphragm

Left and right lower thoracic nodes drain directly to the thoracic duct or via a separate left bronchiomediastinal trunk which joins the thoracic duct in the posterior superior mediastinum. Upper right thoracic nodes drain via the right bronchomediastinal trunk which either drains into the right lymphatic duct or directly into the right subclavian vein.

Note: Normal drainage from breast is to anterior and posterior axillary, infra-clavicular and internal thoracic groups. With pathological blockage from disease the spread can be to opposite side, cervical, peritoneal cavity and liver, and inguinal glands.

3

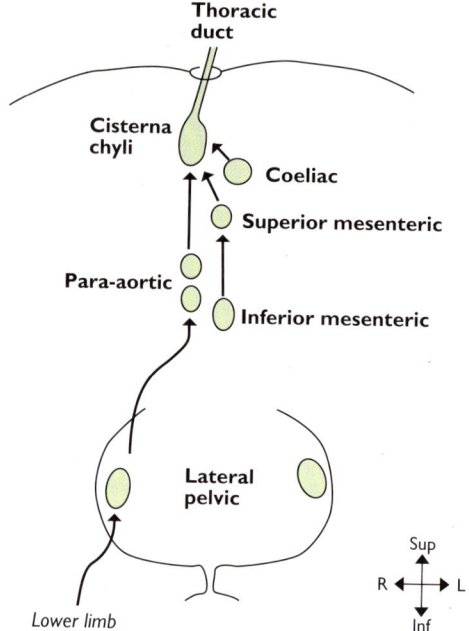

Lymph glands — abdomen

LYMPH GLANDS — ABDOMEN

Coeliac
  Lower third of oesophagus
  Stomach and greater omentum
  First and upper second part of duodenum
  Spleen
  Pancreas
  Liver
  Gallbladder
Superior mesenteric
  Lower second, third and fourth parts of
    duodenum
  Jejunum
  Ileum
  Caecum and appendix
  Ascending colon
  Transverse colon
Inferior mesenteric
  Distal transverse colon
  Descending colon
  Sigmoid colon
  Upper rectum and rectal mucosa to
    dentate line
Para-aortic
  Inferior surface of diaphragm
  Suprarenal gland
  Kidney
  Gonad (plus fallopian tube in female)
  Superior lateral uterus

Ureter
Bare area of liver
Posterior abdominal wall
Lateral pelvic nodes
  Lower rectum and dentate line
  Bladder
  Urethra
  Lower ureter
  Female
    uterus
    cervix
    upper vagina
    clitoris
    labia minora
  Male
    vas deferens
    seminal vesicles
    prostate
    bulk of penis

Endodermal (gut-tube/gut-derived) structures drain to lymph nodes lying along their arteries of supply which are named according to the artery with which they are associated and are not listed individually. Their number and exact position are variable. However, the position and drainage of the highest node groups (listed above) are constant.

3

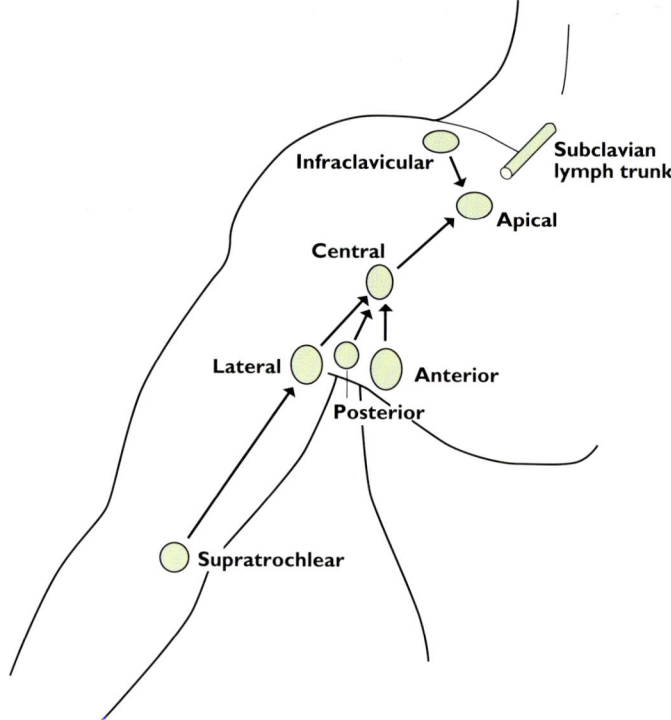

Lymph glands — upper limb

LYMPH GLANDS — UPPER LIMB

Axillary groups
Anterior (pectoral)
  Breast
  Anterior thoracic wall
  Upper anterior abdominal wall
Posterior (subscapular)
  Posterior thoracic wall
  Tail of breast
  Upper posterior abdominal wall

Lateral
  Arm
  Forearm
  Hand
Central
Apical
Supratrochlear
  Skin of anterior forearm and hand
Infraclavicular
  Skin of shoulder
  Skin of lower neck
  Skin of anterior upper thoracic wall
  Breast

3

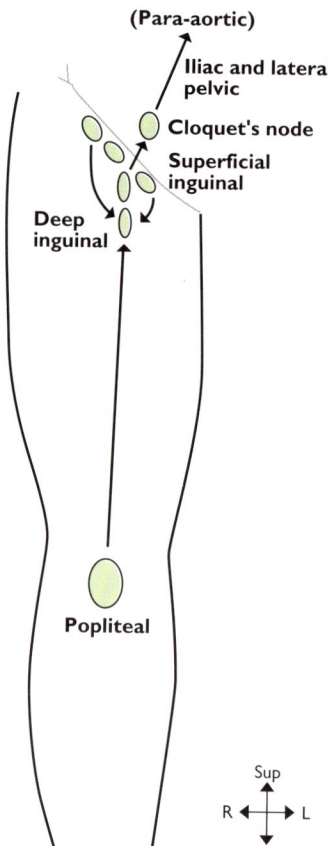

Lymph glands — lower limb

LYMPH GLANDS — LOWER LIMB

Superficial inguinal
  Uterine fundus/skin of penis
  Labia minora/scrotum
  Skin of buttock
  Skin of lower abdominal wall to
    umbilicus
  Skin of thigh, anterior skin of calf and
    dorsum of foot
  Skin of anterior perineum
Deep inguinal
  Anterior perineum

Thigh
Leg
Foot
Popliteal
  Skin of sole of foot
  Skin of posterior calf

Note: Cloquet's node is the highest node of
the lower limb and usually lies within the
femoral canal beneath the inguinal ligament,
medial to the femoral vein and lateral to the
lacunar ligament.

3

4

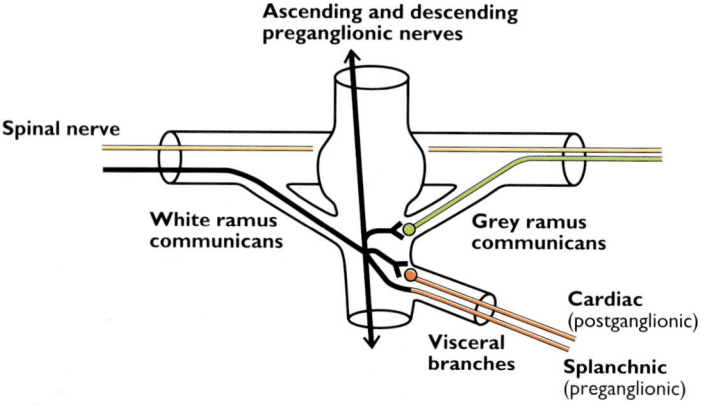

Ascending and descending
preganglionic nerves

Spinal nerve

White ramus
communicans

Grey ramus
communicans

Cardiac
(postganglionic)

Visceral
branches

Splanchnic
(preganglionic)

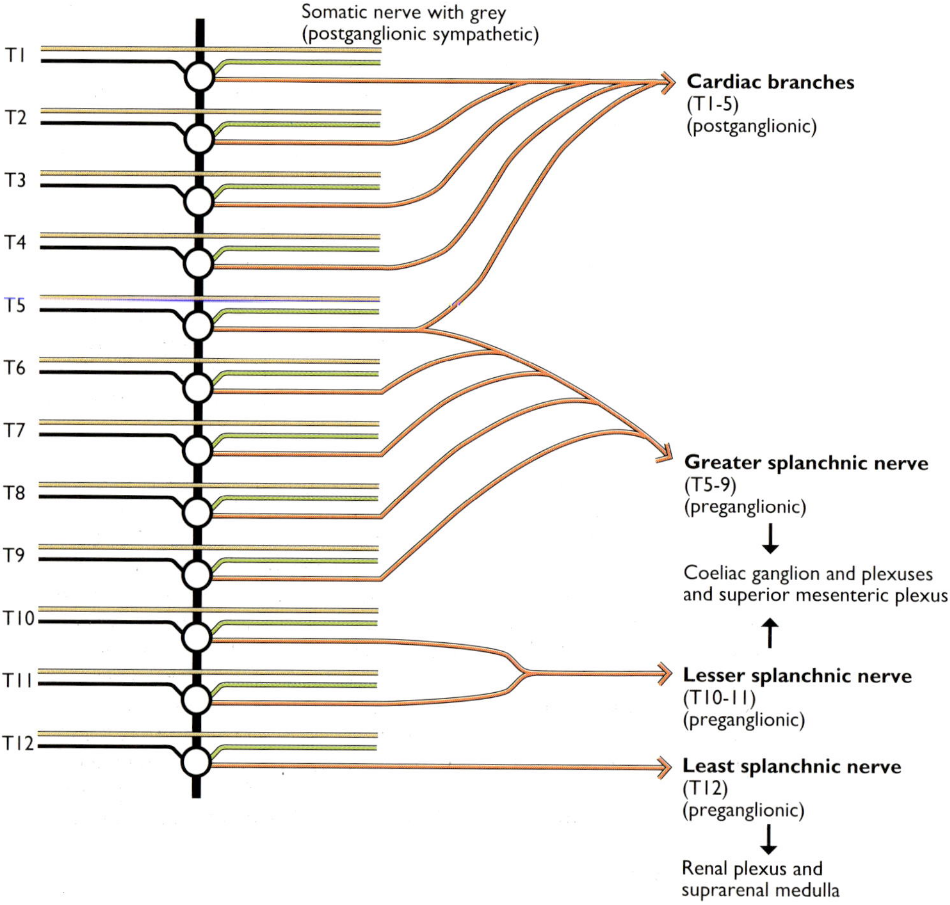

Somatic nerve with grey
(postganglionic sympathetic)

T1

T2

T3

T4

T5

T6

T7

T8

T9

T10

T11

T12

Cardiac branches
(T1-5)
(postganglionic)

Greater splanchnic nerve
(T5-9)
(preganglionic)

↓

Coeliac ganglion and plexuses
and superior mesenteric plexus

↑

Lesser splanchnic nerve
(T10-11)
(preganglionic)

Least splanchnic nerve
(T12)
(preganglionic)

↓

Renal plexus and
suprarenal medulla

Thoracic sympathetics (T1–12)
Note: All splanchnic nerves synapse in collateral ganglia

Sympathetic

The primary motor function of the sympathetic system is to supply vaso-constrictor fibres throughout the body and, more specifically, sudomotor and pilomotor control to the skin via the spinal nerves. In addition it has specific actions such as dilating the pupil. It also has afferent fibres that detect sensation from visceral structures. The following description is an outline of the general plan of the sympathetic nervous system.

Sympathetic outflow from the spinal cord takes place only from preganglionic cell bodies in the lateral horns of T1 to L2. Between these two levels the white (myelinated) rami communicantes emerge with the anterior rami and enter the ganglia of the sympathetic chain. Below this level in the lumbosacral region there is a ganglion at the level of each spinal nerve but the white ramus communicans reaches it by passing down from the T1−L2 region in the sympathetic chain. Above T1 there are three cervical ganglia (superior, middle and inferior) that relay the sympathetic supply via the white rami communicantes from the T1−L2 region (mostly from the upper few thoracic levels) to the cervical somatic nerves and the rest of the head and neck. If the inferior cervical ganglion is fused with the uppermost thoracic ganglion it is termed the 'stellate ganglion'.

Thus, the efferent *preganglionic* white rami (shown in black) have the following alternative pathways on reaching a ganglion in the sympathetic chain:

1 They can synapse with grey (unmyelin-ated) rami communicantes (green) which then supply sympathetic fibres to the spinal nerves (yellow) at the same level.

2 They can pass upwards or downwards to synapse at another level in the manner described in 1.

3 They can pass out of the ganglion as a visceral branch (red) to reach a smaller col-lateral ganglion, such as in the coeliac plexus, which is nearer the organ of its destination where they then synapse to become postganglionic nerves. Some visceral branches, such as the cardiac nerves in the upper thoracic and cervical regions, synapse in the ganglia of the sympathetic chain and are then distributed distally as postgang-lionic fibres (also red).

4 They can synapse in a cervical ganglion and pass out as a postganglionic vascular branch (grey).

5 A few preganglionic fibres reach the suprarenal gland to synapse with the cells of the medulla.

Each ganglion has a somatic and visceral branch but in addition, the three cervical ganglia have a vascular branch which allows for a wider distribution via the arteries than would be possible via cervical somatic nerves alone.

Afferent (sensory) fibres (not shown here) in the sympathetic system travel back via the sympathetic chain to the T1−L2 region where they pass via the white ramus communicans to the posterior root ganglion of the spinal nerve in which their cell bodies are situated. To reach the sympathetic ganglia these afferent nerves travel either on blood vessels, somatic nerves or the sympathetic nerves supplying the various plexuses. Thus, the white rami communi-cantes contain both preganglionic sympath-etic fibres and afferent sympathetic fibres.

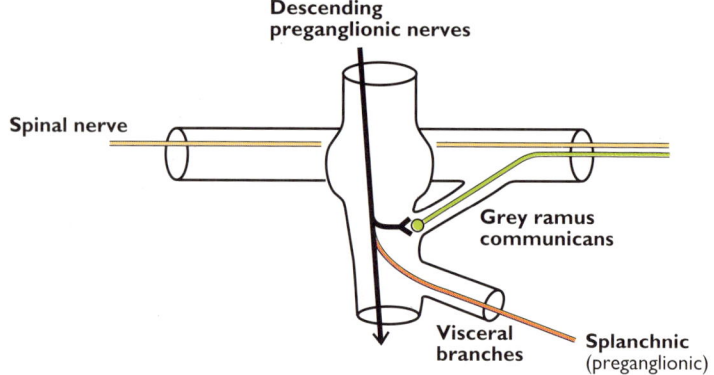

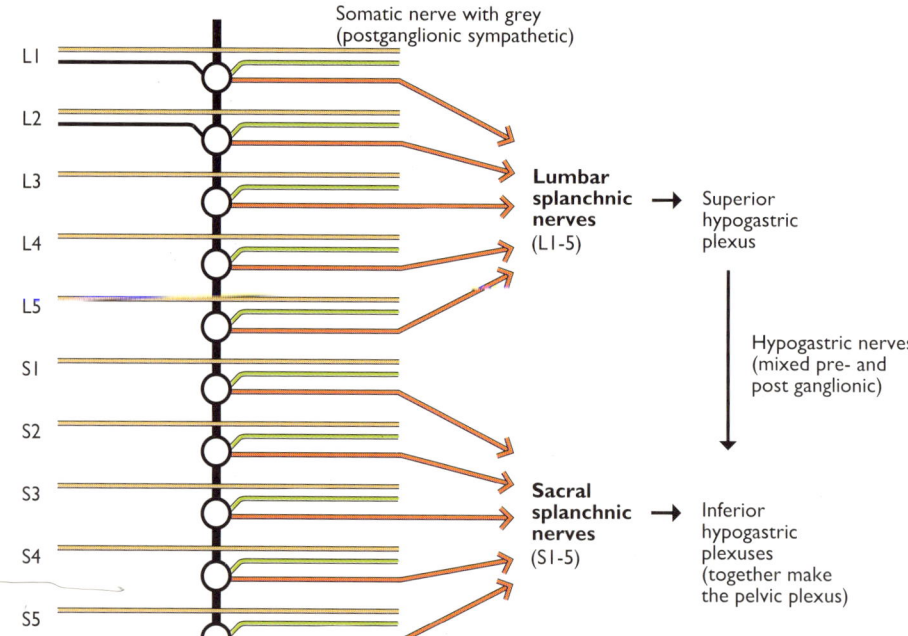

Lumbosacral sympathetics (LI—S5)

Note: LI—2 have white rami communicans. Lumbar and sacral splanchnic nerves are all preganglionic. They synapse in either the superior or inferior hypogastric plexuses

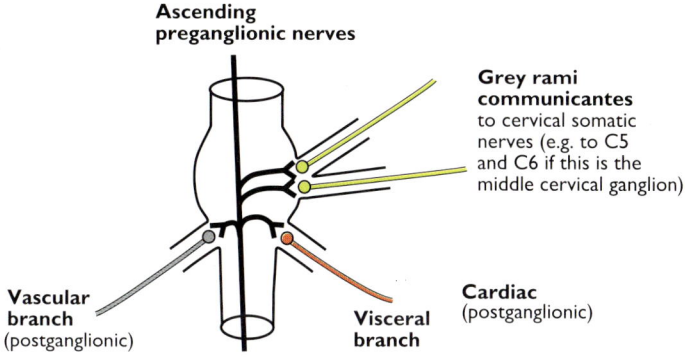

Ascending preganglionic nerves

Grey rami communicantes
to cervical somatic nerves (e.g. to C5 and C6 if this is the middle cervical ganglion)

Vascular branch
(postganglionic)

Visceral branch

Cardiac
(postganglionic)

4

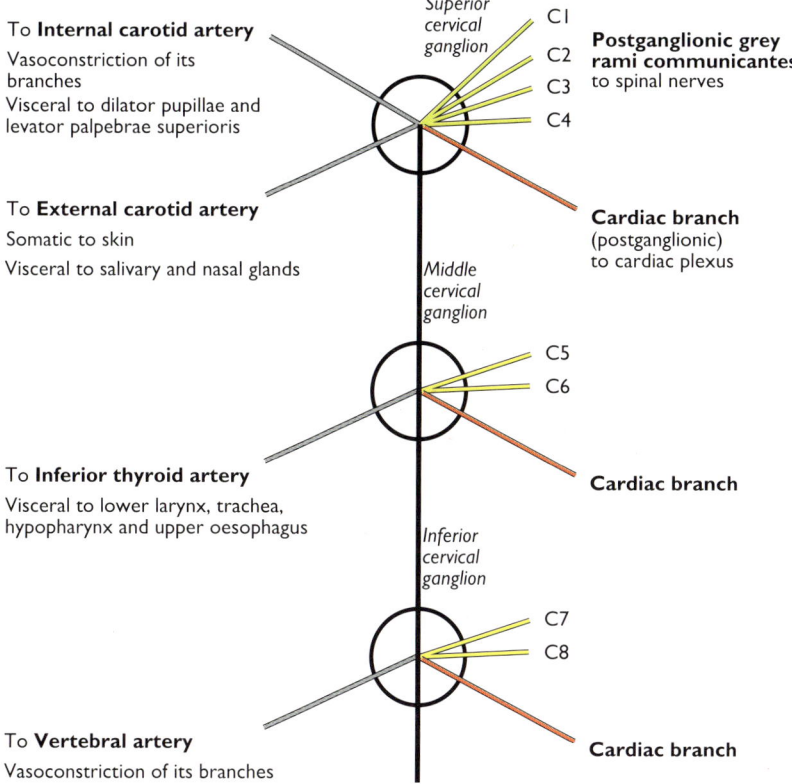

To **Internal carotid artery**

Vasoconstriction of its branches
Visceral to dilator pupillae and levator palpebrae superioris

Superior cervical ganglion

C1
C2
C3
C4

Postganglionic grey rami communicantes
to spinal nerves

To **External carotid artery**

Somatic to skin
Visceral to salivary and nasal glands

Middle cervical ganglion

Cardiac branch
(postganglionic)
to cardiac plexus

C5
C6

To **Inferior thyroid artery**

Visceral to lower larynx, trachea, hypopharynx and upper oesophagus

Inferior cervical ganglion

Cardiac branch

C7
C8

To **Vertebral artery**
Vasoconstriction of its branches

Cardiac branch

Cervical sympathetics (C1−8)

73

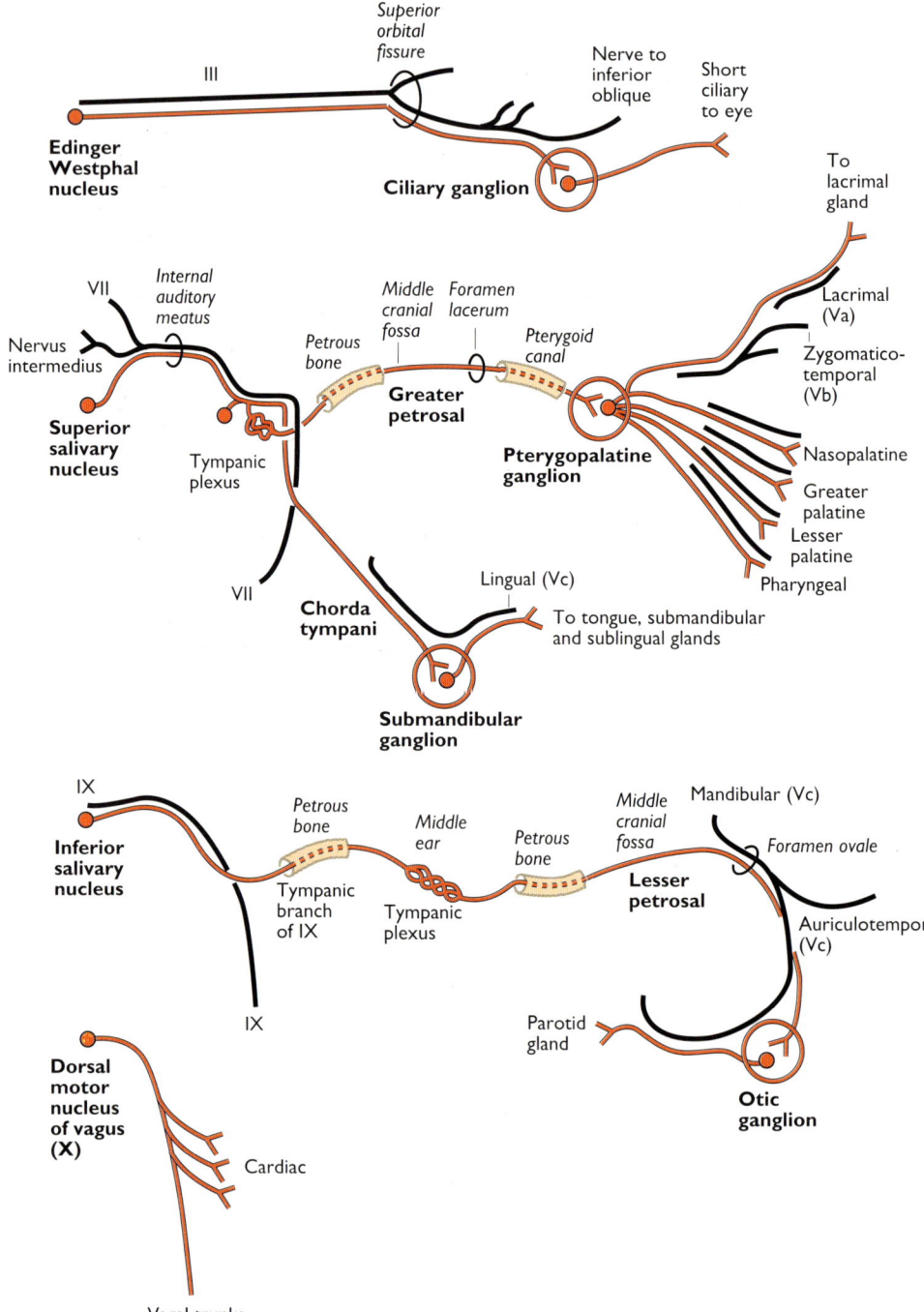

Parasympathetic system
Note: Parasympathetic afferent nerves have been omitted for clarity

Parasympathetic

The parasympathetic system is supplied primarily by four cranial nerves (III, VII, IX and X) with the addition of some outflow from S2, 3 and 4.

In the head and neck the myelinated preganglionic efferent fibres (red) run with the fibres of their relevant cranial nerve (III, VII and IX) to synapse in peripheral named ganglia close to the organ of their destination. They synapse within these ganglia distributing unmyelinated postganglionic fibres to the organ (also red). The fibres may run separately for a period or may associate with other cranial nerve fibres, notably those of the various divisions of the 5th cranial nerve (trigeminal) to reach their final destinations.

The vagal (X) and pelvic parasympathetic fibres (S2, 3 and 4) supplying the viscera below the thoracic inlet always run independently but are distributed to peripheral ganglia where they also synapse. The distribution of these postganglionic fibres is usually very short into the visceral organs of their destination. In *these* nerves there is a much greater degree of ramification of fibres than in those distributed only to the head and neck.

Afferent parasympathetics consist of taste (special visceral sensory) fibres from the tongue, palate and vallecula. In the case of the anterior two thirds of the tongue and palate, the fibres pass through the submandibular and pterygopalatine ganglia respectively without synapsing and their cell bodies are in the geniculate ganglion on the facial nerve (VII). The taste fibres from the posterior third of the tongue are carried in the glossopharyngeal nerve (IX) with the cell bodies in the inferior ganglion on that nerve. Taste from the vallecula is carried by the vagus (X) with the cell bodies in the inferior ganglion of that nerve.

In addition the glossopharyngeal nerve (IX) receives afferent parasympathetic fibres from the carotid body and sinus whose cell bodies are also in the inferior ganglion of that nerve.

Finally, the vagus carries general visceral afferent fibres from the thorax and abdomen, probably to the nucleus solitarius, with the cell bodies also in the inferior vagal ganglion.

4

4

*Note: Apart from cranial nerves 1 and 11,
which clearly arise from specific sensory areas*
*and their fibres pass directly to the brain, a
nomenclature has been used to describe each
nerve as arising centrally in the brain and
passing out to its terminal branches irres-
pective of whether the nerve carries sensory,
motor or a combination of fibres.*

Colour coding in figures
- Somatic motor — black
- Somatic sensory — black
- Special visceral motor (branchial
muscles) — blue
- Special visceral sensory (taste and arterial
receptors) — green
- General visceral motor (parasympath-
etic) — red
- General visceral sensory (parasympath-
etic) — green
- Special senses — black

5

Olfactory nerve (I)

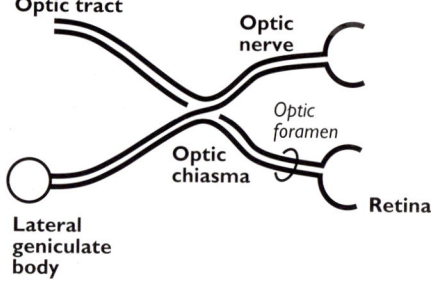

Optic nerve (II)

OLFACTORY NERVE (I)
From: **Olfactory epithelium**
To: **Olfactory cortex**
Contains: **Special sense (smell)**

The olfactory epithelium lines the superior surface of the superior concha, upper medial nasal septum and inferior surface of the cribriform plate of the ethmoid bone. The fibres of the olfactory cells run in the sub-mucosa to pass through the cribriform plate of the ethmoid bone where they synapse in the olfactory bulb which lies on its superior surface. The bulb leads posteriorly to the olfactory tract which lies in the anterior cranial fossa on the inferior surface of the frontal lobe and conveys fibres to the anterior olfactory nucleus (in the posterior aspect of the olfactory bulb), to the prepiri-form cortex, anterior perforating substance and septal areas.

OPTIC NERVE (II)
From: **Retina**
To: **Lateral geniculate body**
Contains: **Special sense (sight)**

The ganglion cells of the retina pass fibres out of the globe of the eye via the optic disc to enter the optic N which passes through the orbit within the dural sheath and within the cone of muscles. The nerve passes through the optic canal in the body of the sphenoid bone into the middle cranial fossa where it lies medial to the anterior clinoid process. The ophthalmic artery lies inferior to it in the canal and runs forwards to pierce the dura around the nerve inferomedially about 1 cm behind the eyeball. The nerve continues posteriorly at first lateral to, then superior to, the sella turcica where it forms the optic chiasma. Fibres from both eyes are distributed to each optic tract with medial retinal fibres (temporal visual fields) crossing to the opposite side. Each tract passes from the posterolateral angle of the chiasma, lying lateral to the pituitary infundibulum, to run lateral to the cerebral peduncle and medial to the uncus of the temporal lobe to reach the lateral geniculate body.

5

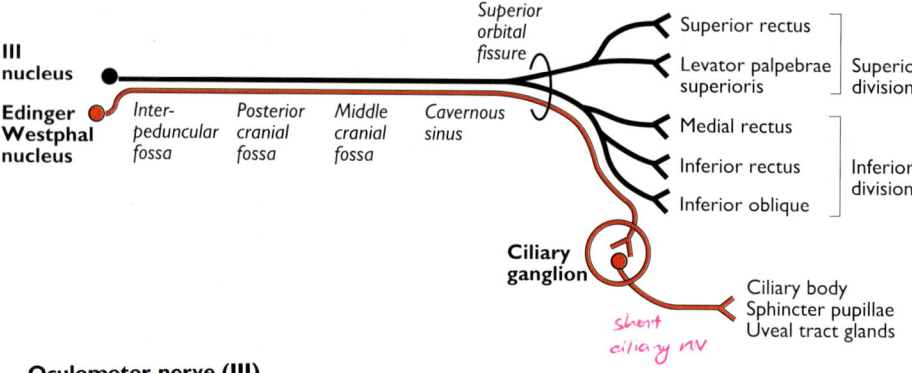

Oculomotor nerve (III)

5

Trochlear nerve (IV)

OCULOMOTOR NERVE (III)

From: Oculomotor nucleus (somatic motor) and Edinger–Westphal nucleus (general visceral motor), ventral to cranial part of aqueduct in midbrain

To: Terminal brs

Contains: Somatic motor & general visceral motor

This nerve emerges medial to the cerebral peduncle in the interpeduncular fossa to reach the middle cranial fossa. It runs forward in close lateral relation to the posterior communicating artery below the margin of the tentorium cerebelli. It pierces the dura lateral to the posterior clinoid process to enter the cavernous sinus lying initially high in its lateral wall. It descends, passing medially over the trochlear N and nasociliary branch of the ophthalmic division of the trigeminal N. It then enters the orbit through the superior orbital fissure within the tendinous ring having divided into superior and inferior divisions at the anterior end of the cavernous sinus. The superior division runs lateral to the optic N on the inferior surface of the superior rectus, passing through this muscle to terminate in levator palpebrae superioris. This division carries sympathetic supply to this muscle from the internal carotid artery. The inferior division divides into terminal branches shortly after passing through the tendinous ring, the nerve to inferior oblique carrying the general visceral motor fibres (parasympathetic) to the ciliary ganglion. This lies posteriorly in the orbit inferolateral to the optic N.

TROCHLEAR NERVE (IV)

From: Trochlear nucleus in floor of aqueduct in dorsal midbrain, level with upper part of inferior colliculus

To: Terminal brs

Contains: Somatic motor

The fibres decussate within the substance of the midbrain to appear on the opposite side. The nerve emerges dorsally and passes lateral to the superior cerebellar peduncle then around the lateral aspect of the mid-brain in the middle cranial fossa to lie just above the superior border of the pons. It runs below the edge of the tentorium cerebelli between the posterior cerebral and the superior cerebellar arteries. It enters the lateral wall of the cavernous sinus where it is crossed medially by the oculomotor N from above down before entering the orbit through the superior orbital fissue superolateral to the tendinous ring. It runs medially above levator palpebrae superioris to terminate as it pierces superior oblique.

5

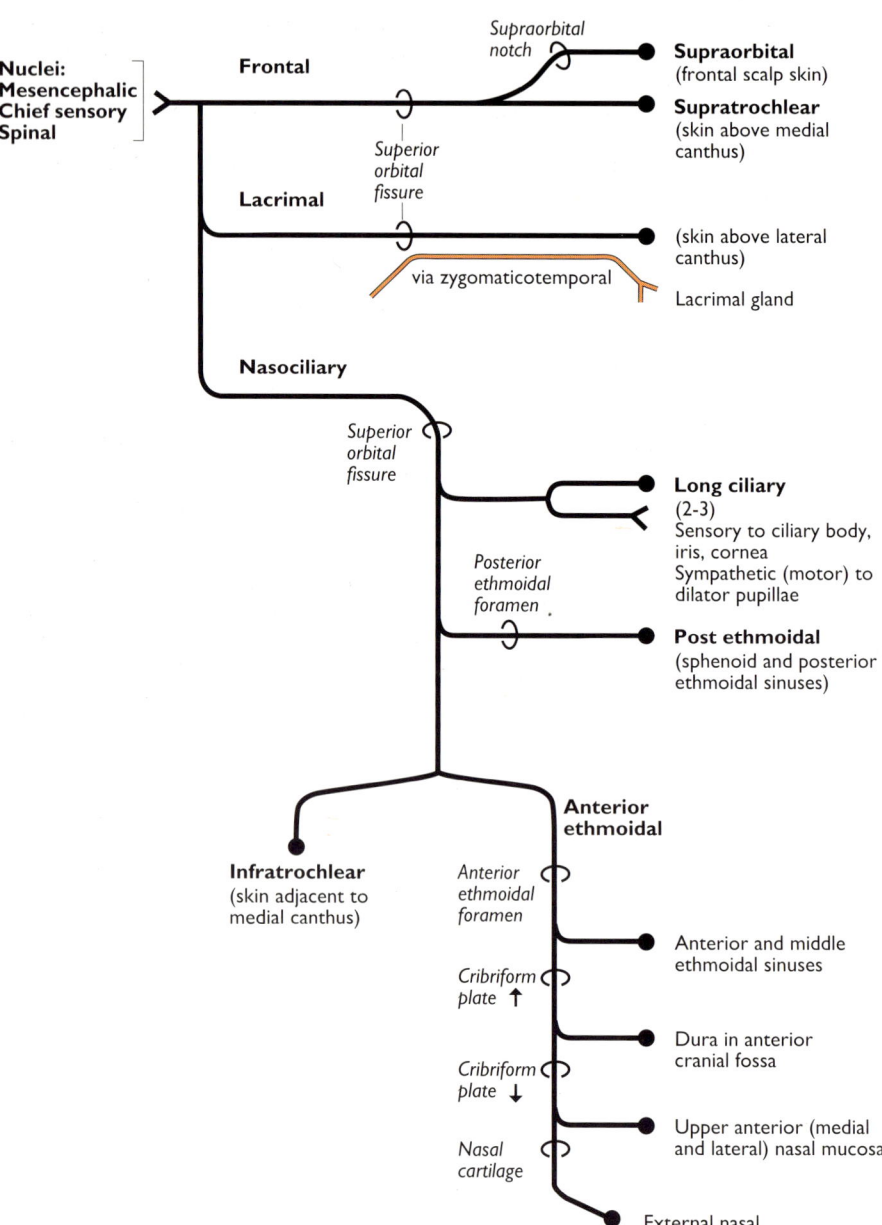

Nuclei:
Mesencephalic
Chief sensory
Spinal

Frontal

Supraorbital notch

● **Supraorbital**
(frontal scalp skin)

Superior orbital fissure

● **Supratrochlear**
(skin above medial canthus)

Lacrimal

● (skin above lateral canthus)

via zygomaticotemporal

Lacrimal gland

Nasociliary

Superior orbital fissure

● **Long ciliary**
(2-3)
Sensory to ciliary body, iris, cornea
Sympathetic (motor) to dilator pupillae

Posterior ethmoidal foramen

● **Post ethmoidal**
(sphenoid and posterior ethmoidal sinuses)

Anterior ethmoidal

Infratrochlear
(skin adjacent to medial canthus)

Anterior ethmoidal foramen

● Anterior and middle ethmoidal sinuses

Cribriform plate ↑

● Dura in anterior cranial fossa

Cribriform plate ↓

● Upper anterior (medial and lateral) nasal mucosa

Nasal cartilage

● External nasal
(skin of nose)

Trigeminal nerve — ophthalmic division (Va)

TRIGEMINAL NERVE —
OPHTHALMIC DIVISION (Va)

From: Terminal nuclei are chief sensory
(touch), mesencephalic (proprioception)
and spinal (pain & temperature). They lie in
pons, midbrain & medulla/upper cervical
cord respectively

To: Terminal brs

Contains: Somatic sensory

The sensory root of the trigeminal N
emerges from the ventral surface of the
upper pons to enter the middle cranial fossa
from where it passes to the trigeminal
ganglion which lies in Meckel's cave, a
prolongation of dura at the apex of the
petrous temporal bone. The ophthalmic
division leaves the trigeminal ganglion and
runs forward in the lateral wall of the
cavernous sinus below the trochlear N and is
crossed medially by the oculomotor N. It
divides into three terminal branches which
pass through the superior orbital fissure
separately. (Note: the mesencephalic nucleus
is unusual in that it receives primary
neurones that do not have the cell bodies in
the ganglion but in the nucleus itself.)

Frontal N. Runs superolateral to the
tendinous ring into the orbit where it
continues forwards and medially above
levator palpebrae superioris, to divide into
terminal branches which leave the orbit over
its superior margin through similarly named
notches.

Lacrimal N. Runs lateral to the tendinous
ring into the orbit. It passes laterally, close to
the periosteum of the orbital plate of the
frontal bone, to terminate in the lacrimal
gland which it supplies together with
adjoining conjunctiva. Some fibres leave the
orbit over its superolateral margin. In its
course it temporarily carries general visceral
motor fibres from the zygomaticotemporal
branch of the maxillary N (Vb) to the
lacrimal gland.

Nasociliary N. Runs within the tendinous
ring between superior and inferior division
of the oculomotor N, crossing superior to
the optic N to lie over the medial rectus. It
leaves the muscular cone giving terminal
branches before running through the
anterior ethmoidal foramen in the ethmoid
bone on the medial orbital wall as the
anterior ethmoidal N. It traverses the
anterior ethmoid sinus to run through its
roof and onto the superior surface of the
cribriform plate beneath the dura of the
anterior cranial fossa. It passes through the
plate again lateral to the crista galli onto the
medial wall of the nose, first on the perpendicular plate of the ethmoid and then on the
inner surface of the nasal bone. It passes into
the skin of the nose beneath the inferior
margin of the nasal bone as its terminal
branch — the external nasal N. The two to
three long ciliary nerves carry sympathetic
fibres to dilate the pupil.

5

5

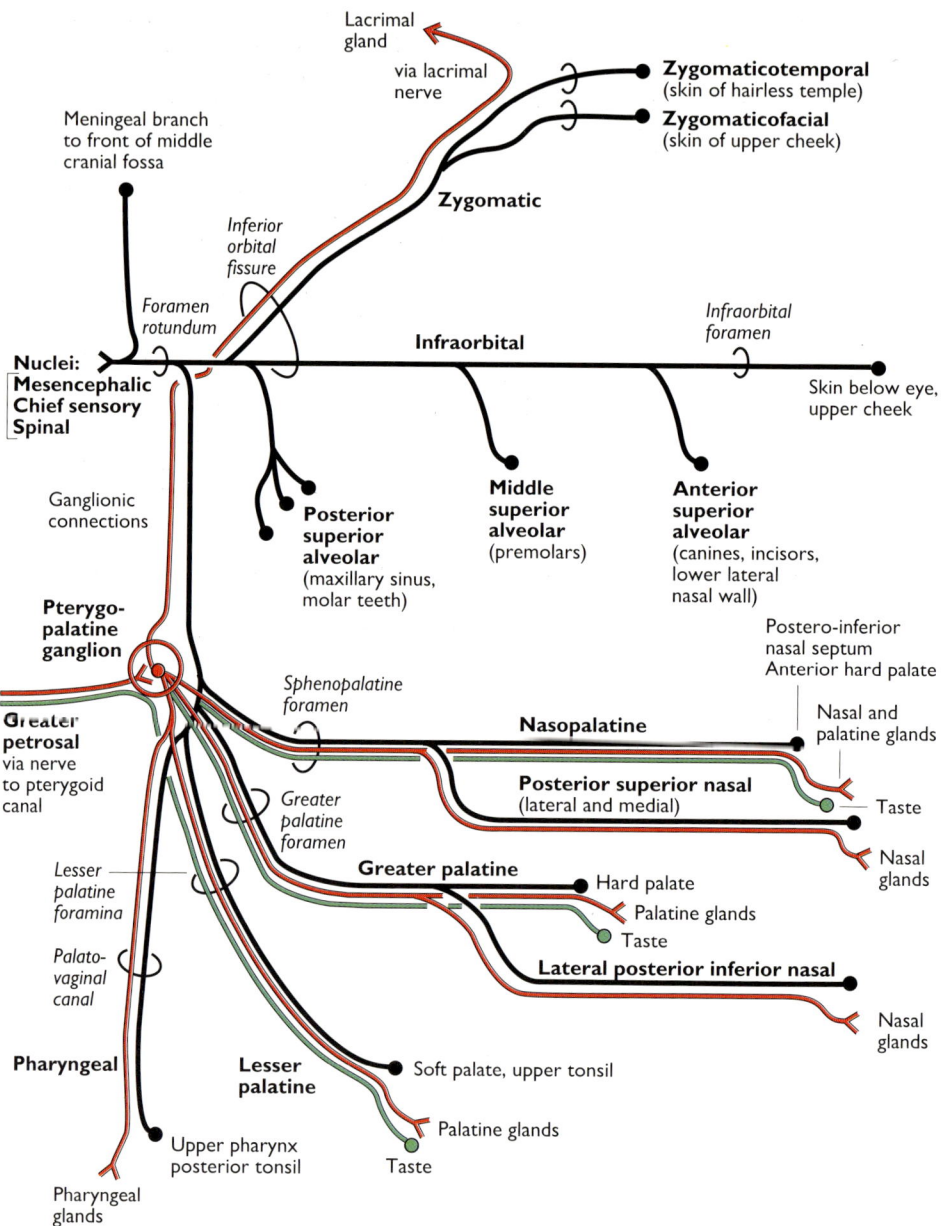

Lacrimal
gland

via lacrimal
nerve

Zygomaticotemporal
(skin of hairless temple)

Zygomaticofacial
(skin of upper cheek)

Zygomatic

Meningeal branch
to front of middle
cranial fossa

*Inferior
orbital
fissure*

*Foramen
rotundum*

*Inferior orbital
fissure*

Infraorbital

*Infraorbital
foramen*

Nuclei:
**Mesencephalic
Chief sensory
Spinal**

Skin below eye,
upper cheek

Ganglionic
connections

**Posterior
superior
alveolar**
(maxillary sinus,
molar teeth)

**Middle
superior
alveolar**
(premolars)

**Anterior
superior
alveolar**
(canines, incisors,
lower lateral
nasal wall)

Postero-inferior
nasal septum
Anterior hard palate

**Pterygo-
palatine
ganglion**

*Sphenopalatine
foramen*

Nasal and
palatine glands

**Greater
petrosal**
via nerve
to pterygoid
canal

Nasopalatine

Posterior superior nasal
(lateral and medial)

Taste

*Greater
palatine
foramen*

Greater palatine

Hard palate

Nasal
glands

*Lesser
palatine
foramina*

Palatine glands

Taste

*Palato-
vaginal
canal*

Lateral posterior inferior nasal

Nasal
glands

Pharyngeal

**Lesser
palatine**

Soft palate, upper tonsil

Upper pharynx
posterior tonsil

Palatine glands

Taste

Pharyngeal
glands

Trigeminal nerve — maxillary division (Vb)

TRIGEMINAL NERVE — MAXILLARY DIVISION (Vb)

From: **Terminal nuclei are chief sensory (touch), mesencephalic (proprioception) and spinal (pain & temperature). They lie in pons, midbrain & medulla/upper cervical cord respectively**

To: **Terminal brs**

Contains: Somatic sensory

(See ophthalmic division for course to the trigeminal ganglion.) The nerve leaves the ganglion to run low down in the lateral wall of the cavernous sinus before passing onto the floor of the middle cranial fossa and it then exits through the foramen rotundum in the greater wing of the sphenoid bone. It runs into the upper pterygopalatine fossa, giving branches via the pterygopalatine ganglion before passing into the orbit via the inferior orbital fissure to become the infra-orbital N. The pterygopalatine ganglion is suspended from the maxillary N by one or two roots and receives general visceral motor fibres from the greater petrosal N in the pterygopalatine fossa which are distributed with the terminal branches as shown.

Infraorbital N. Passes laterally across the posterior aspect of the palatine bone and maxilla to pass through the inferior orbital fissure and to run into the infra-orbital groove in the orbital surface of the maxilla. It terminates in branches in the skin of the face having passed through the infra-orbital foramen which is the completed continuation of the infra-orbital groove.

Zygomatic N. Passes superiorly out of the pterygopalatine fossa through the inferior orbital fissure to run in the lateral orbit outside the cone of muscles. It terminates on the lateral orbital wall as facial and temporal branches which pass through unnamed canals in the zygomatic bone to reach the skin over the zygomatic bone and hairless temple respectively.

Nasopalatine (previously sphenopalatine) N. Passes through the sphenopalatine foramen to enter the posterior upper nasal cavity to end as terminal branches.

Note: Other branches pass through foramina as indicated.

5

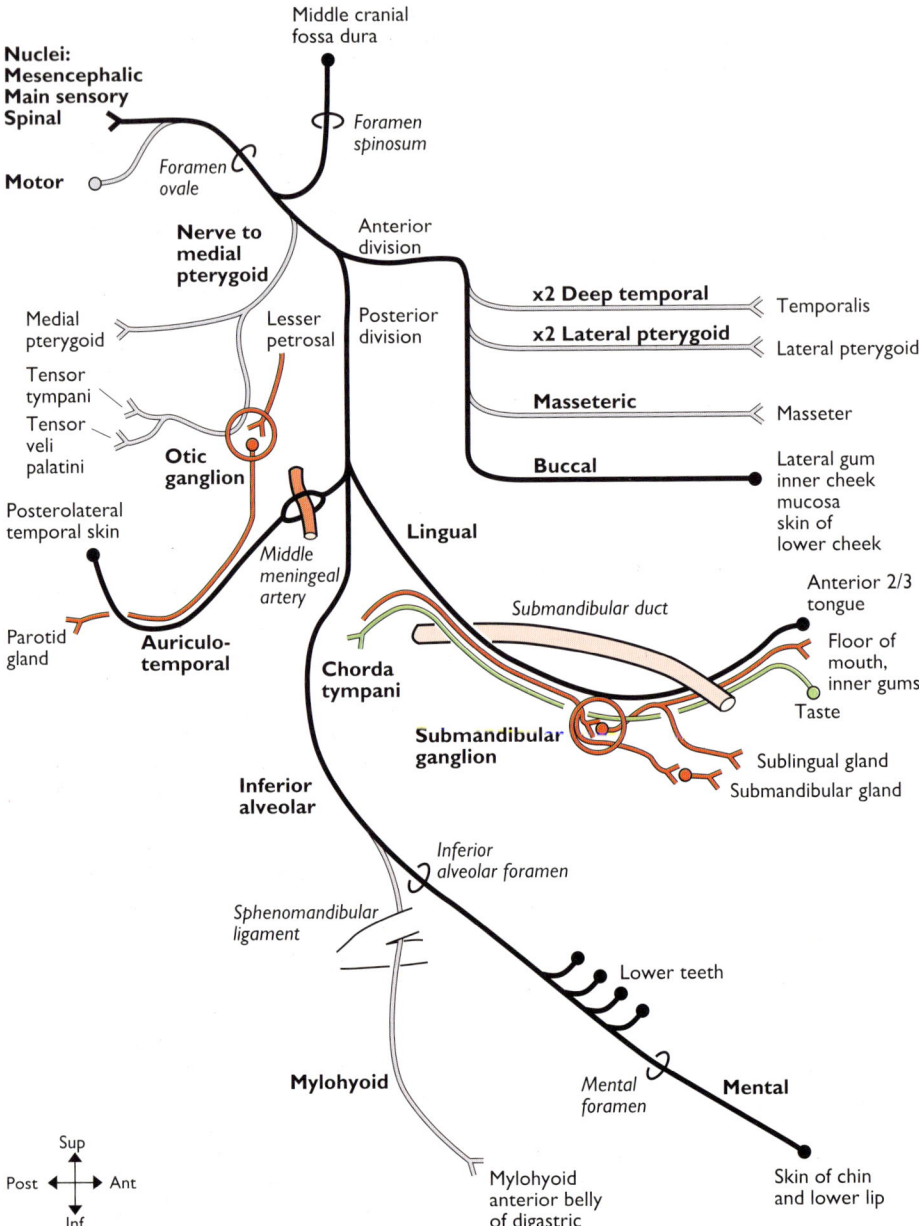

Nuclei:
Mesencephalic
Main sensory
Spinal

Motor

Middle cranial
fossa dura

Foramen
spinosum

Foramen
ovale

Nerve to
medial
pterygoid

Anterior
division

Medial
pterygoid

Lesser
petrosal

Posterior
division

x2 Deep temporal　　Temporalis

x2 Lateral pterygoid　Lateral pterygoid

Masseteric　　Masseter

Buccal　　Lateral gum
inner cheek
mucosa
skin of
lower cheek

Tensor
tympani

Tensor
veli
palatini

Otic
ganglion

Posterolateral
temporal skin

Middle
meningeal
artery

Lingual

Submandibular duct

Anterior 2/3
tongue

Floor of
mouth,
inner gums

Taste

Parotid
gland

Auriculo-
temporal

Chorda
tympani

Submandibular
ganglion

Sublingual gland
Submandibular gland

Inferior
alveolar

Inferior
alveolar foramen

Sphenomandibular
ligament

Lower teeth

Mylohyoid

Mental
foramen

Mental

Sup
Post ← → Ant
Inf

Mylohyoid
anterior belly
of digastric

Skin of chin
and lower lip

Trigeminal nerve — mandibular division (Vc)

TRIGEMINAL NERVE –
MANDIBULAR DIVISION (Vc)

From: Terminal nuclei are chief sensory (touch), mesencephalic (proprioception) and spinal (pain & temperature). They lie in pons, midbrain & medulla/upper cervical cord respectively. Motor nucleus (branchial muscles) is in upper pons

To: Terminal brs

Contains: Somatic sensory & special visceral motor

(See ophthalmic division for course to the trigeminal ganglion.) The smaller motor root leaves the ventral pons anteromedial to the sensory root. The sensory root leaves the ganglion from its lateral part and passes after a short course over the greater wing of the sphenoid bone through the foramen ovale. The motor root passes under the ganglion and unites with the sensory root just beyond the foramen ovale. The nerve that is so formed passes into the infratemporal fossa between tensor veli palatini and lateral pterygoid. It has a short course of 3–4 mm before dividing into anterior and posterior divisions which provide terminal branches.

Anterior division

Deep temporal Ns. Usually two, run above the lateral pterygoid, over the infratemporal crest and on the squamous temporal and greater wing of the sphenoid bones deep to temporalis to supply it. They lie with their associated vessels deeply in the temporal fossa.

Lateral pterygoid Ns. Pierce the muscle directly to supply it.

Masseter N. Runs laterally over the lateral pterygoid and over the mandibular notch to pierce the deep surface of the muscle to supply it.

Buccal N. Runs forward over lateral pterygoid and lies deep to temporalis, the mandible and masseter. It runs over buccinator to supply a small area of overlying ·skin. The nerve is sensory only.

Posterior division

Auriculotemporal N. Passes posteriorly, briefly dividing to encircle the middle meningeal artery, before running between the neck of the mandible and the sphenomandibular ligament. It winds around the neck of the mandible to pass laterally then superiorly lying between the temporomandibular joint and the external auditory meatus deep to the parotid gland. It terminates at the upper border of the gland as branches. The nerve receives general visceral motor fibres from the lesser petrosal N via the otic ganglion which lies suspended from the nerve by two roots close to its origin near the foramen ovale.

Lingual N. Passes forward and inferiorly to lie between lateral pterygoid and tensor veli palatini, then between medial pterygoid and the ramus of the mandible. It lies just beneath the mucous membrane of the mouth posteromedial to the third molar tooth. It passes lateral to styloglossus and hyoglossus and runs at first lateral, then inferior and then medial to the submandibular duct. It terminates over the lateral aspect of the anterior two thirds of the tongue. It is joined 2 cm anterior and inferior to the foramen ovale by general visceral motor and special visceral sensory fibres of the chorda tympani which relay in the submandibular ganglion. This ganglion which is suspended from the nerve by two roots lies on hyoglossus above the submandibular gland.

Inferior alveolar N. Passes deep to lateral pterygoid to lie between the sphenomandibular ligament and the ramus of the mandible before entering the mandible via the inferior alveolar foramen. It terminates as cutaneous branches reappearing through the mental foramen in the anterior body of the mandible. The N to mylohyoid arises just before it enters the inferior alveolar foramen and pierces the sphenomandibular ligament to run in a groove on the medial surface of the body of the mandible below mylohyoid.

5

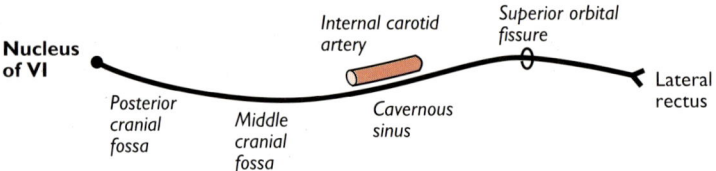

Abducent nerve (VI)

ABDUCENT NERVE (VI)

From: Abducent nucleus in sup part of floor
 of 4th ventricle in lower pons
To: Terminal brs
Contains: Somatic motor

The fibres leave the pons at its lower border above the pyramid of the medulla. It traverses the pontine basal cistern running forwards and superiorly to pierce the dura over the clivus inferolateral to the dorsum sellae. It arches forward directly over the ridge of the petrous temporal bone passing through the medial wall of the inferior petrosal sinus under the petroclinoid ligament and runs onto the medial wall of the cavernous sinus. Here it lies directly lateral to the internal carotid artery before passing into the orbit through the superior orbital fissure within the tendinous ring. It passes forward and laterally to sink into the medial surface of the lateral rectus muscle.

5

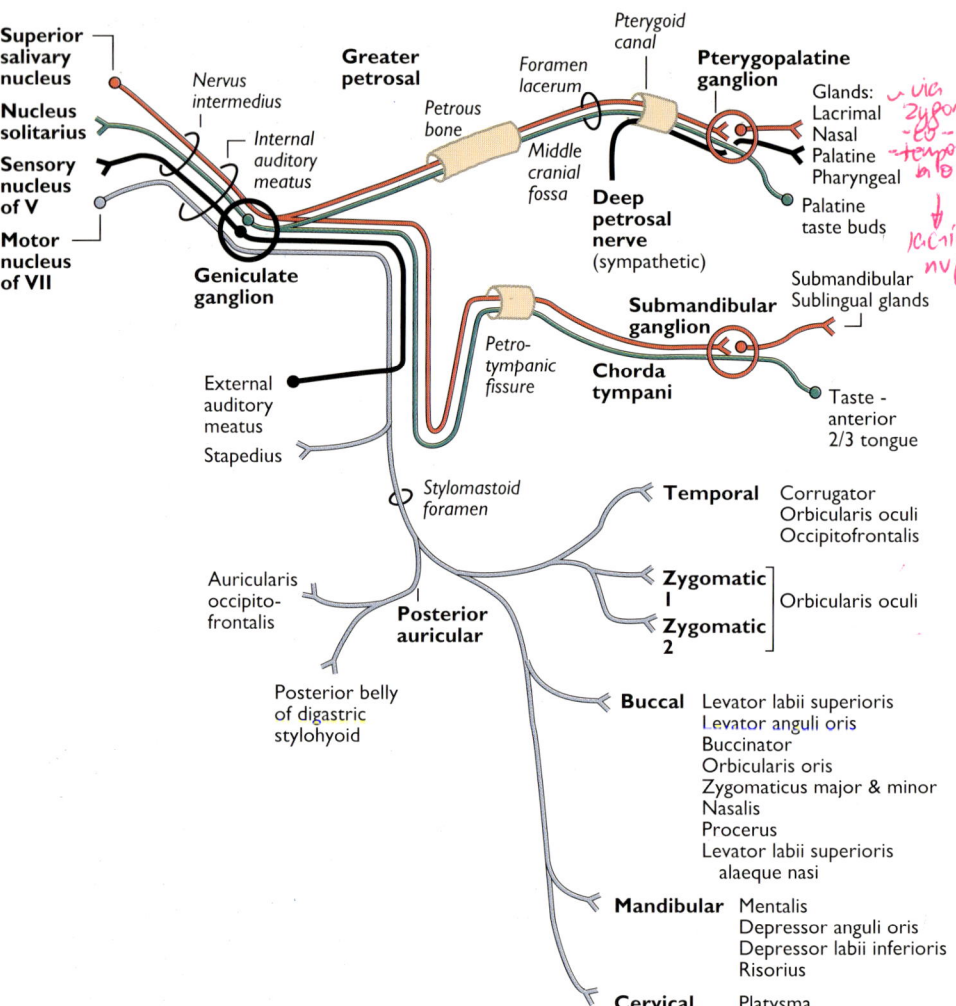

Superior salivary nucleus

Nucleus solitarius

Sensory nucleus of V

Motor nucleus of VII

Nervus intermedius

Internal auditory meatus

Geniculate ganglion

Greater petrosal

Petrous bone

Foramen lacerum

Middle cranial fossa

Pterygoid canal

Deep petrosal nerve (sympathetic)

Pterygopalatine ganglion

Glands:
Lacrimal
Nasal
Palatine
Pharyngeal

Palatine taste buds

Submandibular
Sublingual glands

Submandibular ganglion

Petro-tympanic fissure

Chorda tympani

Taste - anterior 2/3 tongue

External auditory meatus

Stapedius

Stylomastoid foramen

Auricularis occipito-frontalis

Posterior auricular

Posterior belly of digastric stylohyoid

Temporal
Corrugator
Orbicularis oculi
Occipitofrontalis

Zygomatic 1
Zygomatic 2
Orbicularis oculi

Buccal
Levator labii superioris
Levator anguli oris
Buccinator
Orbicularis oris
Zygomaticus major & minor
Nasalis
Procerus
Levator labii superioris alaeque nasi

Mandibular
Mentalis
Depressor anguli oris
Depressor labii inferioris
Risorius

Cervical
Platysma

Facial nerve (VII)

FACIAL NERVE (VII)

From: Facial motor nucleus deep to reticular formation in lower pons. Sup salivary nucleus (general visceral motor) distal to motor nucleus. Gustatory nucleus (taste) in superior end of nucleus solitarius in medulla. Sensory nucleus of V (see trigeminal N)

To: Terminal brs

Contains: Somatic sensory, special visceral sensory & motor, general visceral motor

It leaves the pons at the cerebellopontine angle medial to the vestibulocochlear N (VIII) as two nerve roots — the facial motor root and the nervus intermedius. The nervus intermedius contains special visceral sensory, general visceral motor and somatic sensory fibres which connect with gustatory, superior salivatory and sensory trigeminal nuclei respectively. These two roots pass across the subarachnoid space together to enter the internal auditory meatus and pass laterally along it to reach and enter the facial canal. The two roots then unite and pass laterally onto the medial wall of the middle ear before turning 90 degrees posteriorly at the geniculate ganglion. It continues posteriorly running above the promontory and oval window and below the lateral semicircular canal. Finally the nerve turns 90 degrees inferiorly to run down in the medial wall of the aditus of the mastoid antrum. It leaves the middle ear via the stylomastoid foramen to pass between the mastoid process and the tympanic ring before passing between the deep and superficial portions of the parotid gland. Within the gland it lies superficial to the styloid process, retro-mandibular vein and external carotid artery before dividing into terminal branches which leave the anterior border of the gland.

Greater petrosal N. Arises from the main nerve at the geniculate ganglion and passes medially through the petrous temporal bone to lie in a groove on its anterior surface beneath the temporal lobe and dura of the middle cranial fossa. It runs beneath the trigeminal ganglion, anteromedially across the foramen lacerum. Here it is joined by the deep petrosal N (sympathetic) and becomes the nerve of the pterygoid canal (Vidian's N). It passes through the pterygoid (Vidian's) canal and out into the pterygo-palatine fossa to enter the posterior aspect of the pterygopalatine ganglion. Fibres are then distributed with branches of the maxillary division (Vb).

Chorda tympani. Arises from the facial nerve in the facial canal during its descent from the medial wall of the middle ear. It runs back into the middle ear on the posterior wall before passing anteriorly between the two layers of the flaccid part of the tympanic membrane and over the handle of the malleus. It leaves the middle ear by passing into the petrous temporal bone and emerges via the petrotympanic fissure to pass into the infratemporal fossa medial to the spine of the sphenoid bone which it grooves. It runs antero-inferiorly, deep to lateral pterygoid, to join the lingual branch of the mandibular division (Vc) 2 cm below the skull.

5

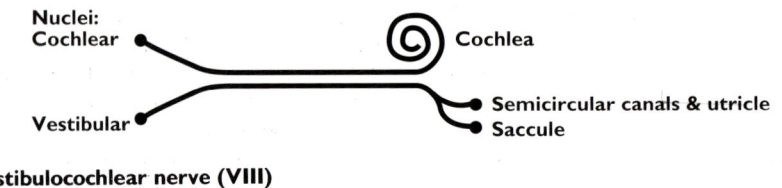

Nuclei:
Cochlear   Cochlea

Vestibular   Semicircular canals & utricle
                             Saccule

Vestibulocochlear nerve (VIII)

5

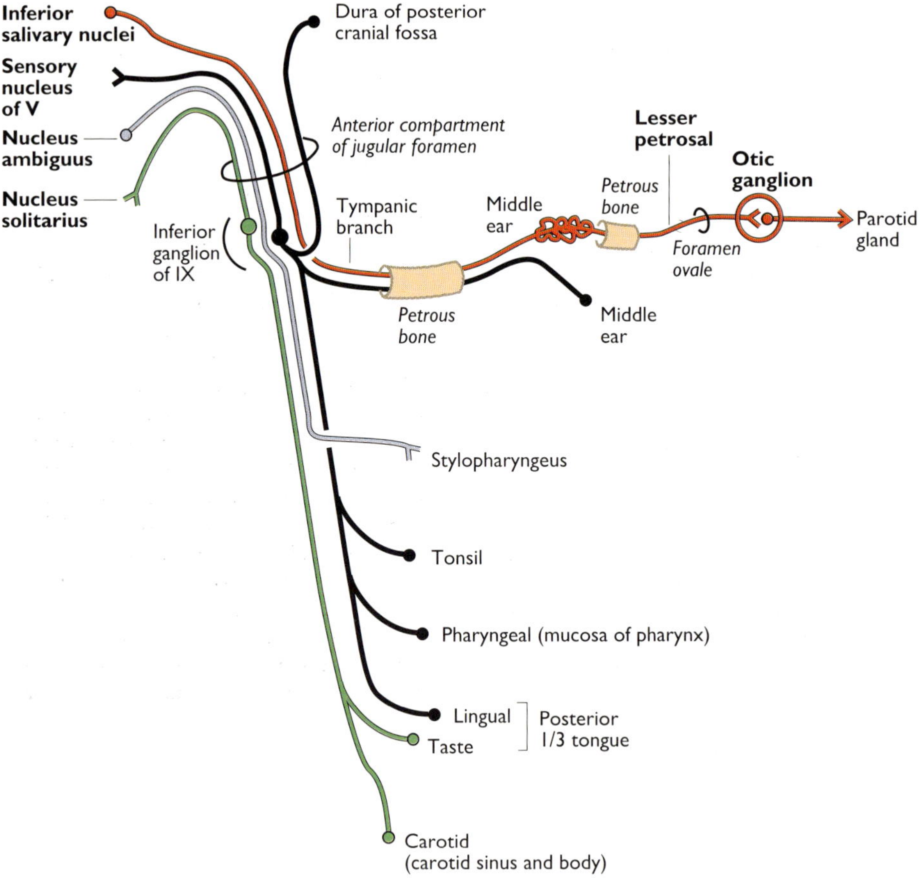

Inferior salivary nuclei ─────── Dura of posterior cranial fossa

Sensory nucleus of V

Nucleus ambiguus ───

Nucleus solitarius ───

Inferior ganglion of IX

Anterior compartment of jugular foramen

Tympanic branch

Middle ear

Petrous bone

Lesser petrosal

Otic ganglion

Foramen ovale

Parotid gland

Petrous bone

Middle ear

Stylopharyngeus

Tonsil

Pharyngeal (mucosa of pharynx)

Lingual
Taste
} Posterior 1/3 tongue

Carotid
(carotid sinus and body)

Glossopharyngeal nerve (IX)

VESTIBULOCOCHLEAR NERVE (VIII)

From: Vestibular & cochlear nuclei in floor of 4th ventricle in pons

To: Inner ear

Contains: Special sense (hearing, balance)

It emerges at the cerebellopontine angle as a single nerve and traverses the subarachnoid space to enter the internal auditory meatus where the cochlear element separates and pierces the temporal bone in its antero-inferior quadrant. The vestibular element divides into upper and lower divisions to pierce the temporal bone in its postero-superior and postero-inferior quadrants. The cochlear N runs in the cochlear modiolus to end in terminal connections. The upper vestibular division runs to supply the semicircular canals and the utricle, the lower division the saccule.

....

GLOSSOPHARYNGEAL NERVE (IX)

From: Sensory nucleus of V (common sensation — see trigeminal N). Nucleus solitarius (taste — medulla). Nucleus ambiguus (motor to branchial muscle — medulla). Inf salivary nucleus (secreto-motor — lower pons)

To: Terminal brs

Contains: Somatic sensory, general & special visceral motor, general & special visceral sensory

The fibres leave the medulla as three or four rootlets lying posterior to the olive. They rapidly fuse into one nerve which passes anterolaterally into the anterior compartment of the jugular foramen (between petrous temporal and occipital bones). It passes medial to the inferior petrosal sinus which separates it from the vagus and accessory Ns and runs anteriorly out of the compartment. It forms the glossopharyngeal ganglia below the compartment as it passes between internal jugular vein and internal carotid artery. It passes inferolaterally looping around the upper border of stylopharyngeus, runs deep to hyoglossus and terminates in lingual and pharyngeal branches.

Carotid N. Arises just below the ganglia and runs down closely adherent to the internal carotid artery within the carotid sheath to reach the carotid sinus and carotid body.

Lesser petrosal N. The tympanic branch (Jacobson's N) of the glossopharyngeal N arises just below the ganglia to pass via the petrous temporal bone into the middle ear. It mingles with parasympathetic fibres of the facial N (VII) and sympathetics from the internal carotid artery over the promontory on the medial wall to form the lesser petrosal N. This nerve then leaves the middle ear via the medial side of the roof and passes through the petrous temporal bone into the middle cranial fossa. Here it lies beneath the dura to run forward before passing through the foramen ovale and synapsing below in the otic ganglion in the infratemporal fossa. From here it is distributed with the auriculo-temporal branch of the mandibular division (Vc).

5

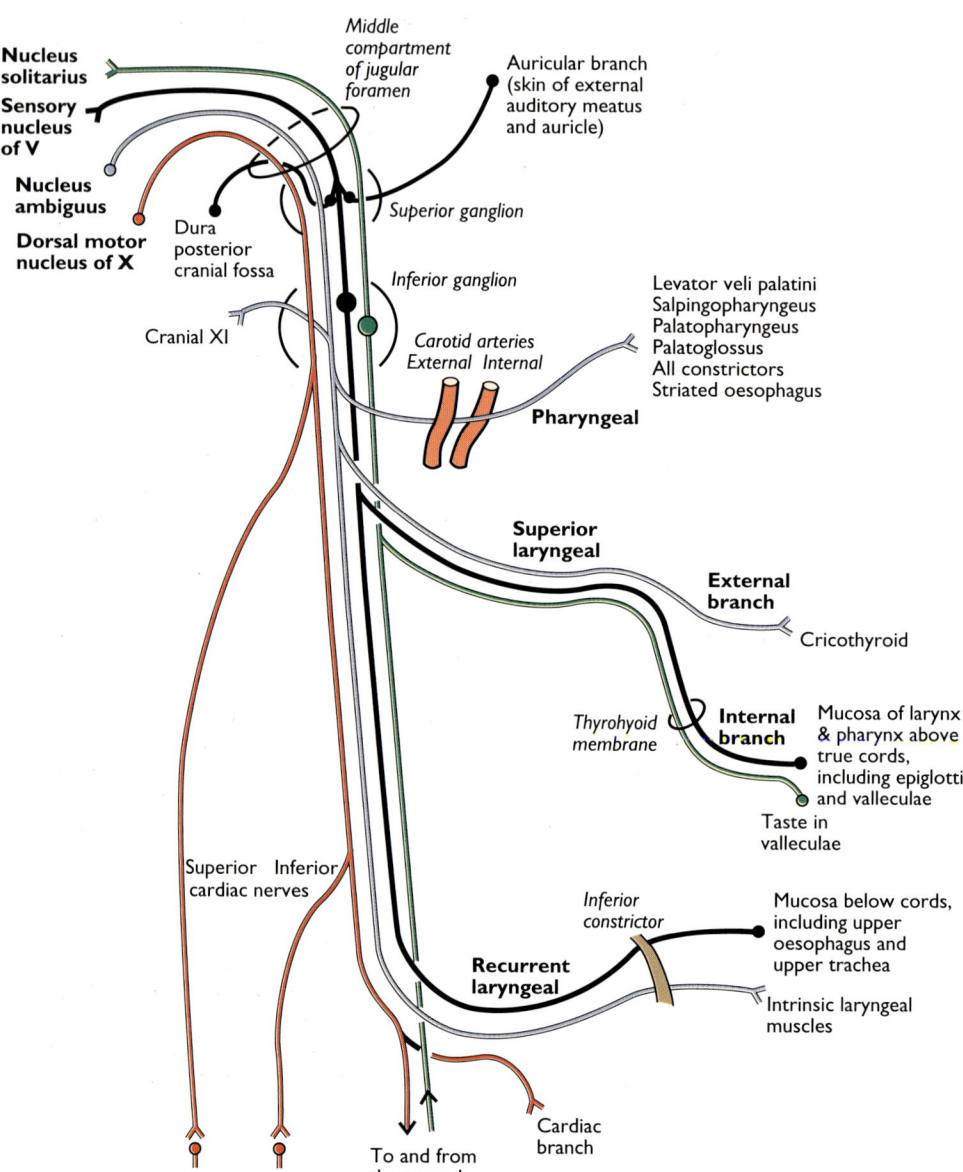

Nucleus solitarius

Sensory nucleus of V

Nucleus ambiguus

Dorsal motor nucleus of X

Middle compartment of jugular foramen

Auricular branch (skin of external auditory meatus and auricle)

Superior ganglion

Dura posterior cranial fossa

Inferior ganglion

Cranial XI

Carotid arteries
External Internal

Levator veli palatini
Salpingopharyngeus
Palatopharyngeus
Palatoglossus
All constrictors
Striated oesophagus

Pharyngeal

Superior laryngeal

External branch

Cricothyroid

Thyrohyoid membrane

Internal branch

Mucosa of larynx & pharynx above true cords, including epiglottis and valleculae

Taste in valleculae

Superior Inferior cardiac nerves

Inferior constrictor

Mucosa below cords, including upper oesophagus and upper trachea

Recurrent laryngeal

Intrinsic laryngeal muscles

Cardiac plexuses

To and from thorax and abdomen

Cardiac branch

Vagus nerve (X)

VAGUS NERVE (X)

From: Dorsal motor nucleus of vagus
(general visceral motor — lower medulla).
Nucleus ambiguus (branchial motor —
medulla). Nucleus solitarius (taste & general visceral sensory — medulla). Sensory
nucleus of V (common sensation — see
trigeminal N)

To: Terminal brs

Contains: Somatic sensory, general & special
visceral sensory, general & special visceral
motor

The fibres emerge from the medulla as a
series of rootlets posterior to the olive
between the glossopharyngeal and cranial
accessory rootlets. These form into a single
nerve that passes into the middle compartment of the jugular foramen. Below
the foramen it forms superior and inferior
ganglia before being joined by the cranial
part of the accessory N. It passes vertically
down within the carotid sheath closely
related to the internal carotid artery and
lying between it and the internal jugular
vein.

Pharyngeal branch. Passes from the vagus at
the inferior ganglion running between
internal and external carotid arteries to
reach the lateral wall of the pharynx. These
fibres are mostly from the cranial part of the
accessory N.

Superior laryngeal N. Passes from the
inferior ganglion running steeply down
anteriorly, lying posterior and then medial
to the internal carotid artery. It pierces the
carotid sheath to run on the wall of the
pharynx to the level of the greater cornu of
the hyoid bone where it divides.

Internal branch. Runs down anteriorly onto
the thyrohyoid membrane which it pierces at
the level of the vallecula, and is then distributed as terminal branches to the mucous membrane of the larynx down to the
vocal folds.

External branch. Runs down over the
inferior constrictor accompanied by the
superior thyroid artery to reach the
cricothyroid muscle.

Recurrent laryngeal N. In the neck the two
sides follow the same course ascending in the
tracheo-oesophageal groove. As the nerve
passes medial to the lateral lobe of the
thyroid gland it is intimately related to the
inferior thyroid artery. It passes beneath the
inferior border of cricopharyngeus (inferior
constrictor) to terminate within the submucosa of the larynx. The nerve on the
right originates from the vagus anterior to
the subclavian artery around which it hooks
posteriorly before running medially to
ascend in the tracheo-oesophageal groove.
The nerve on the left originates from the
vagus inferolateral to the arch of the aorta
passing inferior to the arch and posterior to
the ligamentum arteriosum. It runs to the
right of the arch as it passes posteriorly over
the left side of the trachea to reach the
tracheo-oesophageal groove.

Cardiac Ns. The upper branch arises below
the inferior ganglion and the lower branch
arises in the root of the neck. On the right
they pass down anterior to the brachiocephalic artery and on the left over the aortic
arch to terminate in the cardiac plexuses.

5

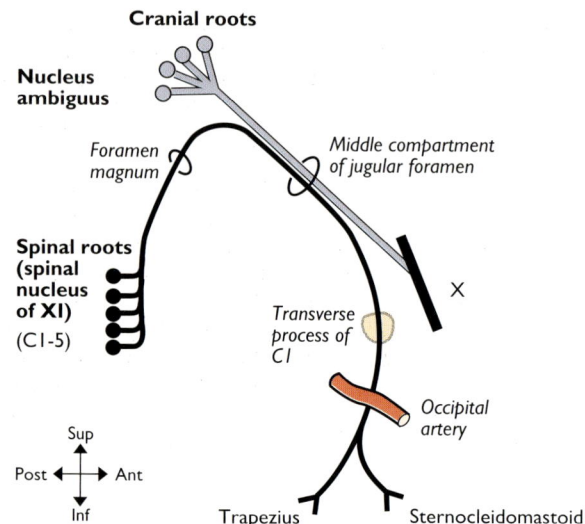

Cranial roots

Nucleus ambiguus

Foramen magnum

Middle compartment of jugular foramen

Spinal roots (spinal nucleus of XI) (C1–5)

X

Transverse process of C1

Occipital artery

Sup

Post ← → Ant

Inf

Trapezius   Sternocleidomastoid

Accessory nerve (XI)

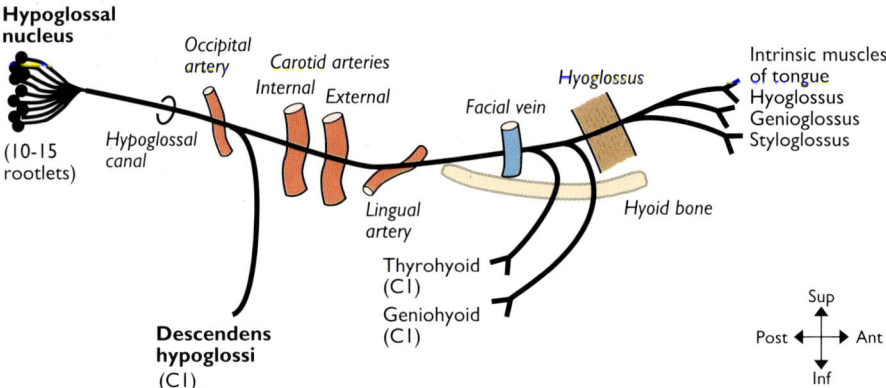

Hypoglossal nucleus

Occipital artery

Carotid arteries

Internal   External

Facial vein

Hyoglossus

Intrinsic muscles of tongue
Hyoglossus
Genioglossus
Styloglossus

(10–15 rootlets)

Hypoglossal canal

Lingual artery

Hyoid bone

Thyrohyoid (C1)

Geniohyoid (C1)

Descendens hypoglossi (C1)

Sup

Post ← → Ant

Inf

Hypoglossal nerve (XII)

ACCESSORY NERVE (XI)

From: **Cranial root from nucleus ambiguus (branchial motor — medulla). Spinal root from spinal nuclei (C1—C5)**
To: Terminal brs
Contains: Somatic motor (spinal), special visceral motor (cranial)

The fibres of the cranial root emerge from the medulla as four to six rootlets posterior to the olive immediately below those of the vagus to fuse into a single nerve. They are joined by the spinal root as it ascends via the foramen magnum (see cervical plexus, pp. 100—101). The nerve passes out of the posterior cranial fossa through the middle compartment of the jugular foramen posterior to the vagus and anterior to the internal jugular vein. Inferior to the foramen the cranial element passes inferomedially to fuse with the vagus to which it adds its complement of special visceral motor fibres.

Spinal root. Passes posterolaterally, usually posterior to the internal jugular vein, over the lateral mass of the atlas (C1) and deep to the occipital artery to enter the deep surface of sternocleidomastoid. It traverses the posterior triangle of the neck from one third of the way down the posterior border of sternocleidomastoid to one third of the way up the anterior border of trapezius where it terminates.

HYPOGLOSSAL NERVE (XII)

From: **Hypoglossal nucleus in floor of 4th ventricle in medulla**
To: Terminal brs
Contains: Somatic motor

Its fibres pass out of the anterolateral surface of the medulla between the olive and pyramid as a series of 10—15 rootlets. These fuse to form two roots which pass posterior to the vertebral artery as they run into the hypoglossal canal where they themselves fuse. The nerve runs out of the canal anteriorly, lateral to the occipital, internal carotid, external carotid and lingual arteries before passing over the apex of the greater cornu of the hyoid bone. It then runs anteriorly, looping over hyoglossus, deep to mylohyoid, to end in terminal branches beneath the submandibular gland.

5

6: PERIPHERAL NERVES

6

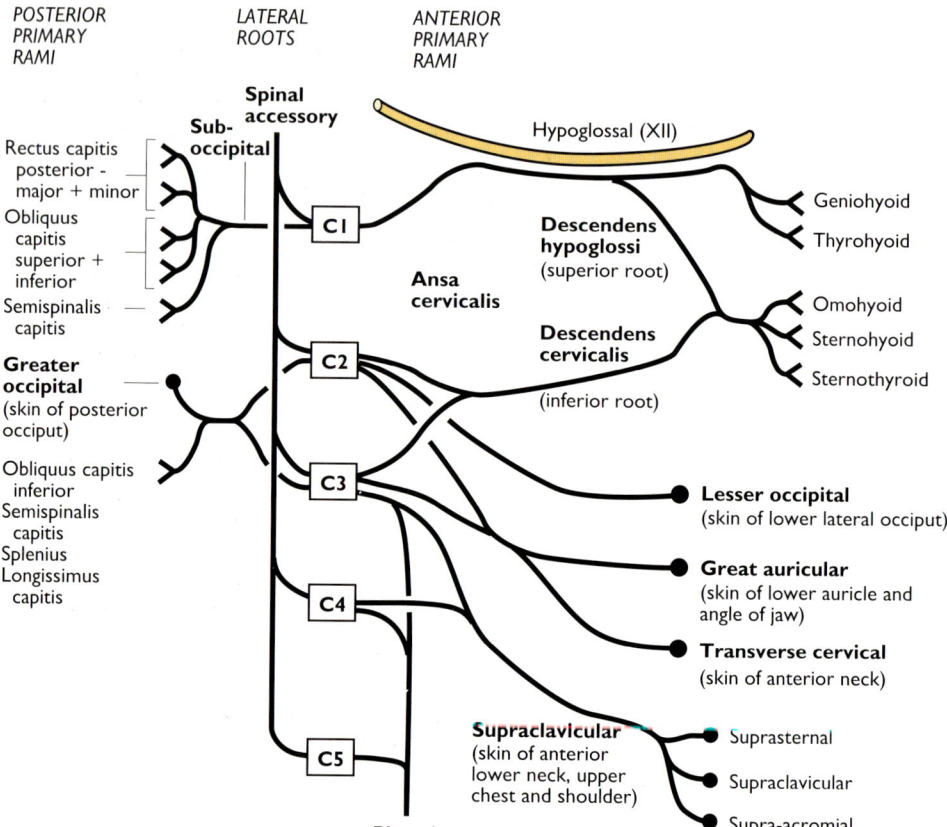

Cervical plexus (C1,2,3,4,5)

CERVICAL PLEXUS (C1,2,3,4,5)
From: C1,2,3,4,5 Ns
To: Ns as shown

It arises mostly from the anterior primary rami deep between scalenus medius and scalenus anterior at the level of C1−C4 vertebrae and is covered by prevertebral fascia lying deep to sternocleidomastoid. The cutaneous branches pierce the prevertebral fascia and run into the posterior triangle of the neck where they pierce the investing layer of the deep cervical fascia to terminate in subcutaneous Ns.

Ansa cervicalis (C1−C3). Upper root (anterior primary rami C1) — passes directly to the hypoglossal N (XII) between rectus capitis anterior and lateralis. It leaves the hypoglossal N lateral to the occipital artery and runs anterior to the internal and common carotid arteries where it joins the lower root. Lower root (anterior primary rami C2,3) — passes laterally around the internal jugular vein having pierced the prevertebral fascia at the level of C2/3. It runs forwards and anteriorly as a long loop to meet with the upper root anterior to the common carotid artery.

Suboccipital N (posterior primary ramus of C1). Emerges through the dura to run beneath the vertebral artery closely applied to the posterior arch of the atlas (C1). It pierces the posterior atlanto-occipital membrane between obliquus capitis superior and rectus capitis posterior major to terminate in muscular branches in the suboccipital triangle.

Greater occipital N (posterior primary rami of C2,3). Emerges from the posterior spinal dura at the intervertebral foramen and passes posteriorly over the transverse process of the axis (C2) below obliquus capitis inferior. It then winds around this muscle to ascend deep to semispinalis piercing it and trapezius near to their insertions into the superior nuchal line. It terminates as cutaneous branches running in the scalp with the occipital artery.

Spinal accessory N (XIs) (lateral roots C1−C5) is formed from the unique lateral roots of C1−C5 and ascends within the subarachnoid space lateral to the cord to pass through the foramen magnum posterior to the vertebral artery to meet with the cranial root.

Phrenic N (see pp. 112−113).

6

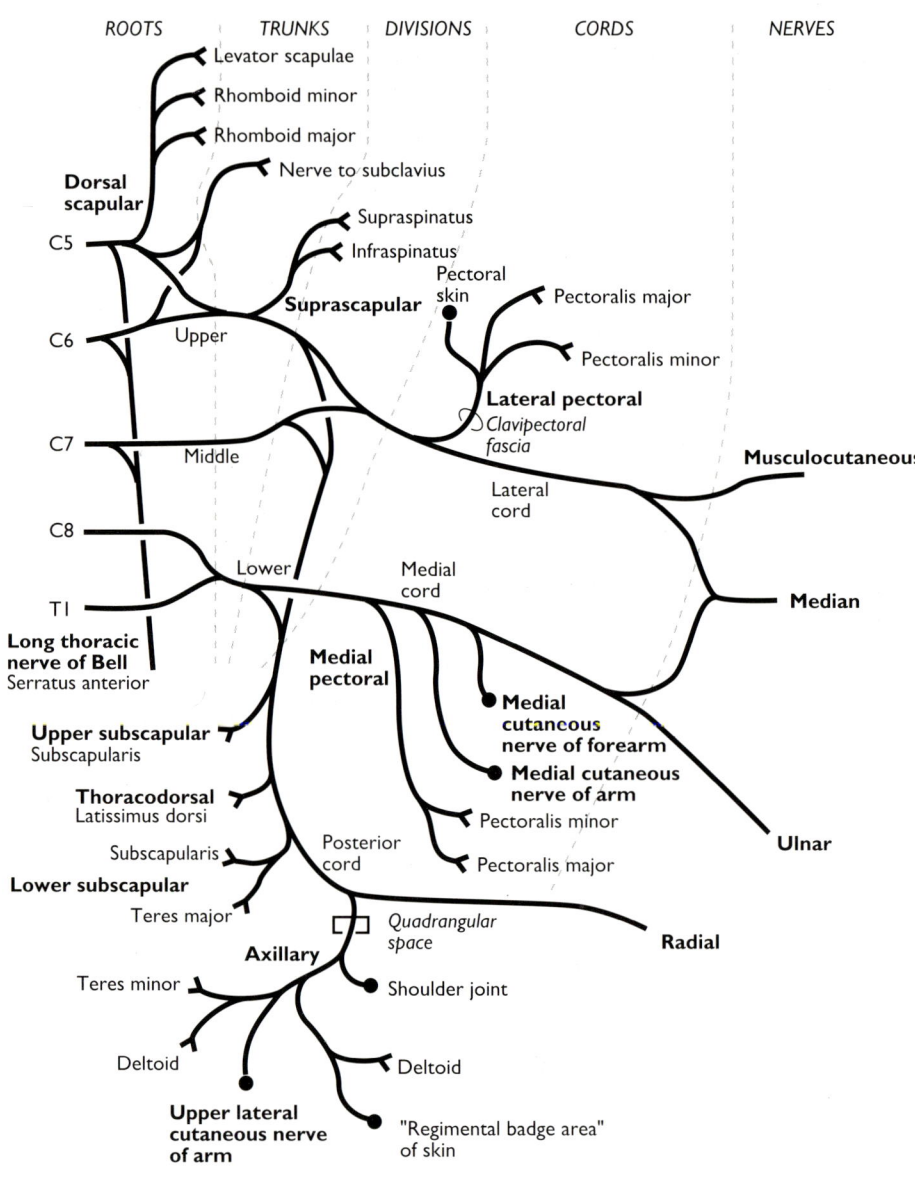

Brachial plexus (C5,6,7,8,T1)

BRACHIAL PLEXUS (C5,6,7,8,T1)
From: Ant primary rami of C5,6,7,8,T1
To: Musculocutaneous, median, ulnar &
    radial Ns

It emerges as five roots lying anterior to
scalenus medius and posterior to scalenus
anterior. The trunks lie in the base of the
posterior triangle of the neck, where they are
palpable, and pass over the 1st rib posterior
to the third part of subclavian artery to
descend to lie behind the clavicle. The
divisions form behind the middle third of the
clavicle lying on the upper fibres of serratus
anterior and around the axillary artery, as
they form the cords. The cords lie in the
axilla related to the second part of the
axillary artery lying medial, lateral and
posterior as their names indicate and
posterior to pectoralis minor. Terminal
nerves are formed around the third part of
the axillary artery posterior to the lower
fibres of pectoralis major.

Axillary N (C5,6). Arises posterior to the
third part of axillary artery. It runs posteriorly on subscapularis to pass through
the quadrangular space with the posterior
circumflex humeral artery. It is intimately
related to the medial aspect of the surgical
neck of the humerus running laterally to end
in anterior and posterior divisions deep to
deltoid.

Thoracodorsal N (C5,6,7). Runs with the
subscapular artery down the medial scapular
border over teres major and into latissimus
dorsi.

Long thoracic N of Bell (C5,6,7). Descends
posterior to the trunks of the plexus and the
first part of the axillary artery to lie on the
lateral aspect of serratus anterior on the
medial axillary wall.

Suprascapular N (C5,6). Arises in the
posterior triangle of the neck, passes
posterolaterally deep to trapezius and
omohyoid and runs through the suprascapular notch into the supraspinous fossa.
It descends laterally around the scapular
spine into the infraspinous fossa.

6

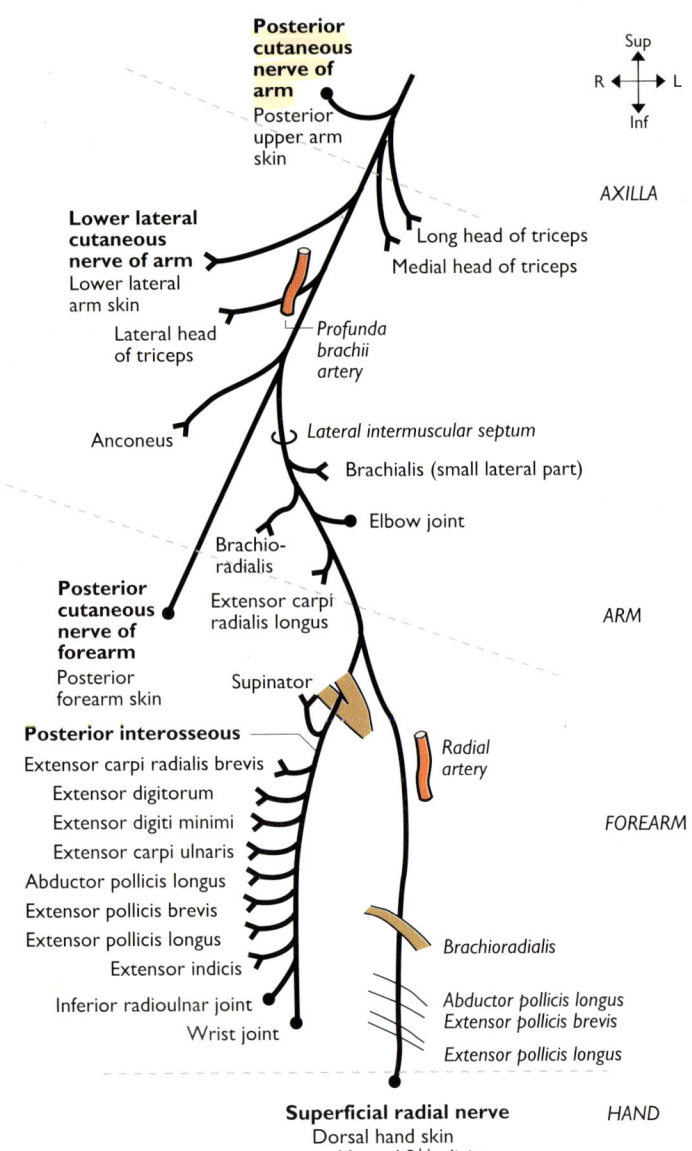

Posterior cutaneous nerve of arm
Posterior upper arm skin

Sup
R ← → L
Inf

AXILLA

Lower lateral cutaneous nerve of arm
Lower lateral arm skin

Long head of triceps
Medial head of triceps

Lateral head of triceps

Profunda brachii artery

Anconeus

Lateral intermuscular septum

Brachialis (small lateral part)

Elbow joint

Brachio-radialis

Posterior cutaneous nerve of forearm
Posterior forearm skin

Extensor carpi radialis longus

ARM

Supinator

Posterior interosseous
Extensor carpi radialis brevis
Extensor digitorum
Extensor digiti minimi
Extensor carpi ulnaris
Abductor pollicis longus
Extensor pollicis brevis
Extensor pollicis longus
Extensor indicis
Inferior radioulnar joint
Wrist joint

Radial artery

FOREARM

Brachioradialis

Abductor pollicis longus
Extensor pollicis brevis
Extensor pollicis longus

Superficial radial nerve
Dorsal hand skin
and lateral $3^{1}/_{2}$ digits

HAND

Radial nerve (C5,6,7,8,T1)

RADIAL NERVE (C5,6,7,8,T1)
From: Post cord of brachial plexus
To: Terminal brs

It arises as the continuation of the posterior cord and descends posterior to the axillary and brachial arteries crossing the tendons of latissimus dorsi and teres major to run with the profunda brachii artery between the long and medial heads of triceps to pass through the lateral triangular space. It gives off the posterior cutaneous N of arm before leaving the axilla. It then runs over the spiral line of the humerus between medial and lateral heads of triceps giving muscular and cutaneous branches, and pierces the lateral intermuscular septum at the mid point of the humerus to reach the anterior compartment. Here it lies deep to the upper fibres of brachialis then brachioradialis, before entering the lateral cubital fossa. It divides into terminal branches over the lateral epicondyle.

Superficial terminal branch. Runs over supinator, pronator teres and flexor digitorum superficialis and lies under brachioradialis running with the radial artery on its medial aspect from one third of the way down the forearm. It passes deep to the tendon of brachioradialis proximal to the radial styloid and over the tendons of the snuff box where it terminates as cutaneous branches to the dorsum of the hand.

Posterior interosseous N. Passes between the two heads of supinator three fingers' breadth below radial head passing into the posterior compartment where it breaks up into terminal muscular branches in the plane between the deep and superficial muscles in this compartment.

6

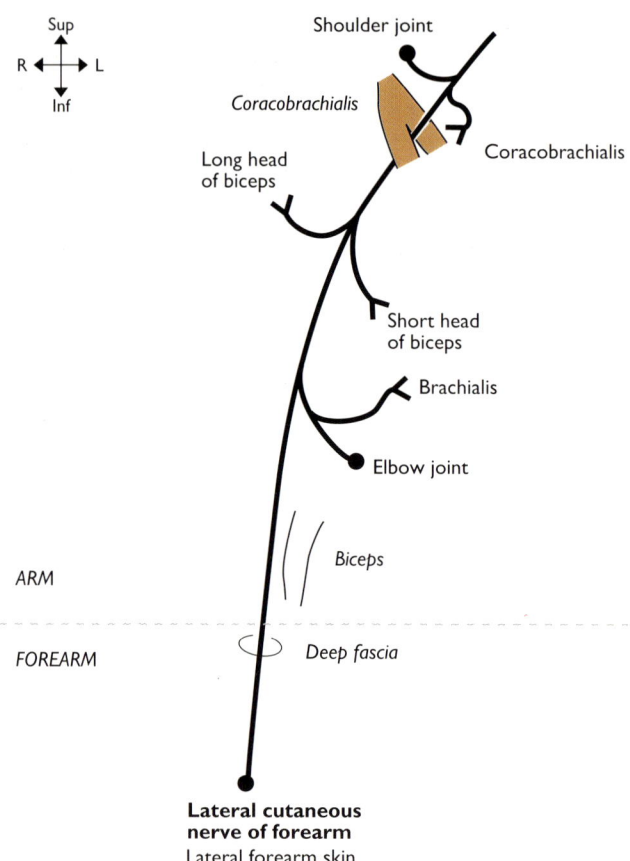

6

Musculocutaneous nerve (C5,6,7)

MUSCULOCUTANEOUS NERVE (C5,6,7)

From: Lat cord
To: Terminal brs

It arises obliquely behind the lower fibres of pectoralis minor lying lateral to the axillary artery and passes laterally between the two conjoined heads of coracobrachialis. It runs laterally downwards between biceps and brachialis, usually adherent to the deep surface of biceps. The terminal branch is the lateral cutaneous N of forearm.

Lateral cutaneous N of forearm. Emerges lateral to the tendon of biceps in the cubital fossa, piercing the deep fascia just below the elbow and descends over the lateral aspect of the forearm to terminate in the skin over the radial artery at the wrist.

6

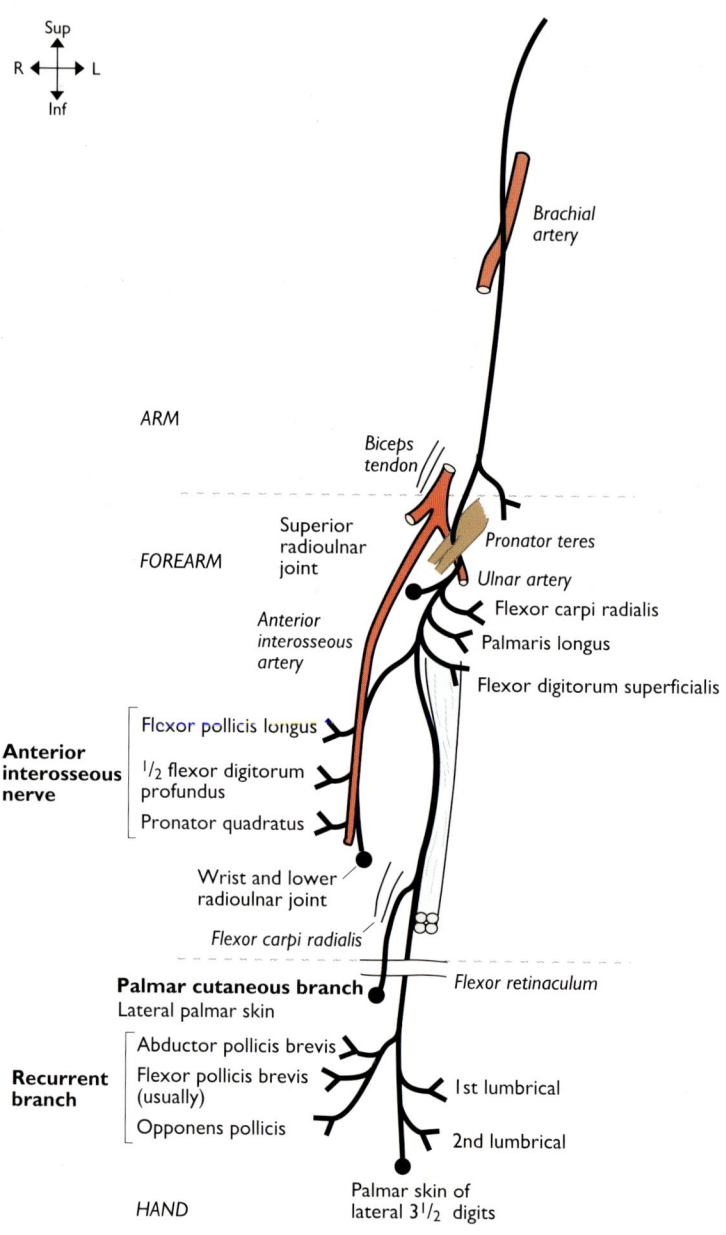

Sup

R ←→ L

Inf

Brachial artery

ARM

Biceps tendon

Superior radioulnar joint

Pronator teres

FOREARM

Ulnar artery

Flexor carpi radialis

Anterior interosseous artery

Palmaris longus

Flexor digitorum superficialis

Anterior interosseous nerve

Flexor pollicis longus

$^1/_2$ flexor digitorum profundus

Pronator quadratus

Wrist and lower radioulnar joint

Flexor carpi radialis

Palmar cutaneous branch
Lateral palmar skin

Flexor retinaculum

Recurrent branch

Abductor pollicis brevis

Flexor pollicis brevis (usually)

Opponens pollicis

1st lumbrical

2nd lumbrical

Palmar skin of lateral $3^1/_2$ digits

HAND

Median nerve (C6,7,8,T1)

MEDIAN NERVE (C6,7,8,T1)
From: Med & lat cords
To: Terminal brs

It is formed in the lower axilla by two roots (heads) which clasp the axillary artery. The nerve initially lies anterior to the axillary artery and then lateral to it and subsequently lateral to the brachial artery. The median N then crosses the brachial artery, usually anteriorly, at the level of the mid humerus, to lie medial to the artery in the cubital fossa. It lies first on coracobrachialis and then brachialis. It passes beneath the bicipital aponeurosis at the elbow leaving the cubital fossa between the two heads of pronator teres before crossing superficial to the ulnar artery and giving its anterior interosseous branch below this. It lies applied to the deep surface of flexor digitorum superficialis on flexor digitorum profundus. It emerges from the lateral side of flexor digitorum superficialis about 5 cm proximal to the wrist where it gives its palmar cutaneous branch and then passes deep to the flexor retinaculum between the tendons of flexor digitorum superficialis and flexor carpi radialis. In the carpal tunnel it divides into terminal branches — recurrent (muscular) branch and palmar digital Ns.

Anterior interosseous N. Arises just below the two heads of pronator teres to run on the interosseous membrane between and covered by flexor digitorum profundus and flexor pollicis longus ending beneath pronator quadratus.

Recurrent branch of median N. Runs out of the carpal tunnel over the distal border of flexor retinaculum onto flexor pollicis brevis to terminate by passing into the thenar eminence.

6

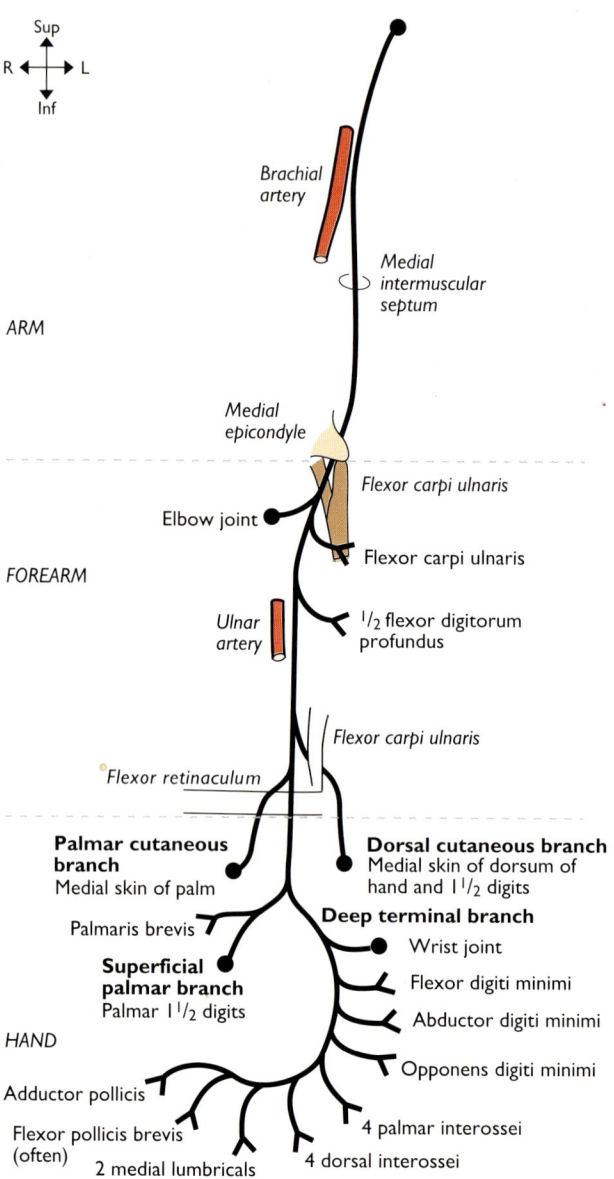

Sup

R ← → L

Inf

Brachial artery

Medial intermuscular septum

ARM

Medial epicondyle

Flexor carpi ulnaris

Elbow joint

Flexor carpi ulnaris

FOREARM

Ulnar artery

$^1/_2$ flexor digitorum profundus

Flexor carpi ulnaris

Flexor retinaculum

Palmar cutaneous branch
Medial skin of palm

Dorsal cutaneous branch
Medial skin of dorsum of hand and $1\,^1/_2$ digits

Deep terminal branch

Palmaris brevis

Wrist joint

Superficial palmar branch
Palmar $1\,^1/_2$ digits

Flexor digiti minimi

Abductor digiti minimi

HAND

Opponens digiti minimi

Adductor pollicis

Flexor pollicis brevis (often)

4 palmar interossei

2 medial lumbricals

4 dorsal interossei

Ulnar nerve (C8,T1)

6

ULNAR NERVE (C8,T1)
From: Med cord of brachial plexus
To: Terminal brs

It arises medial to the axillary artery and continues medial to the brachial artery lying on coracobrachialis to the mid point of the humerus where it leaves the anterior compartment by passing posteriorly through the medial intermuscular septum with the superior ulnar collateral artery. It lies between the intermuscular septum and the medial head of triceps passing posterior to the medial humeral epicondyle and enters the forearm between the two heads of flexor carpi ulnaris. It then lies medial to the coronoid process of the ulna, runs deep to flexor carpi ulnaris and on flexor digitorum profundus, with the ulnar artery on its lateral side from one third of the way down the forearm. It lies lateral to the tendon of flexor carpi ulnaris at the wrist and then passes superficial to the flexor retinaculum to divide into terminal branches at the pisiform bone.

Superficial terminal branch — lies superficial in the palm terminating as terminal digital Ns.

Deep terminal branch – passes through the hypothenar eminence between flexor digiti minimi and abductor digiti minimi grooving the hook of the hamate and runs with the deep palmar arch, deep to the flexor tendons to terminate in adductor pollicis.

Dorsal cutaneous branch — arises 5 cm proximal to the wrist, passes deep to flexor carpi ulnaris onto the medial aspect of the dorsum of the hand where it terminates as cutaneous Ns.

6

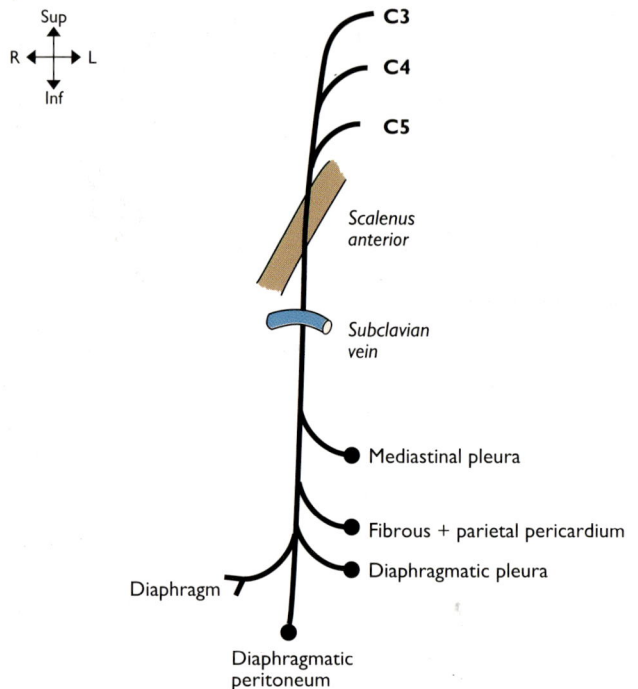

Phrenic nerve (C3,4,5)

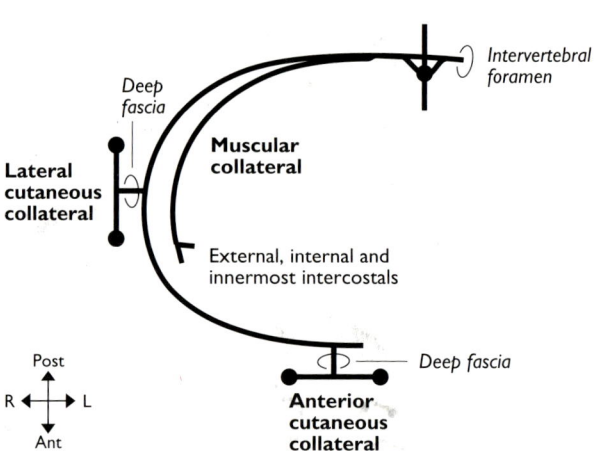

Intercostal nerve

PHRENIC NERVE (C3,4,5)
From: Ant primary rami of C3,4,5
To: Terminal brs

It arises deep between scalenus medius and scalenus anterior and runs over the lateral border of scalenus anterior beneath the prevertebral fascia. It runs on scalenus anterior from lateral to medial lying lateral to the ascending cervical artery and it passes behind the suprascapular and transverse cervical arteries as it does so. It runs over the anterior part of the dome of the pleura to enter the mediastinum posterior to the subclavian vein where right and left nerves take different courses. Right — spirals forwards to lie lateral to the right brachiocephalic vein and continues on the lateral surface of the superior vena cava, right atrium and inferior vena cava, lying within the fibrous pericardium. It traverses the diaphragm via the caval orifice. Left — descends usually anterior to the left internal thoracic artery lying lateral to the left common carotid artery. It runs down over the aortic arch crossing anterior to the left vagus before running anterior to the left pulmonary artery. It then runs lateral to the left auricle and left ventricle within the fibrous pericardium to traverse the diaphragm in isolation via the muscular portion of the diaphragm to the left of the central tendon.

An accessory phrenic N (C5) arising from the N to subclavius may join the phrenic N near the 1st rib.

INTERCOSTAL NERVE (TYPICAL)
From: Ant primary rami of thoracic N
To: Terminal brs

It emerges from the intervertebral foramen (giving off the posterior primary ramus as it does so) to pass between the pleura and the inner muscle layer anterior to the transverse process where it connects via the grey and white rami communicantes with the thoracic sympathetic chain. It passes posterior to the intercostal artery to lie below it as it runs in the subcostal groove in the plane between internal and innermost muscle layers. The muscular collateral branch arises before the posterior angle of the rib and runs in the same plane but lies at the level of the upper border of the rib below.

T1. Has no lateral or anterior cutaneous branches.

T7–11. Run behind the costal margins in their anterior course to lie in the same muscle plane in the abdomen. At their anterior limit they pass deep to rectus abdominis in the rectus sheath and pierce both of these structures to give terminal anterior cutaneous branches. These also supply rectus abdominis.

T12. Passes below the 12th rib as the subcostal N having similar branches to those above.

6

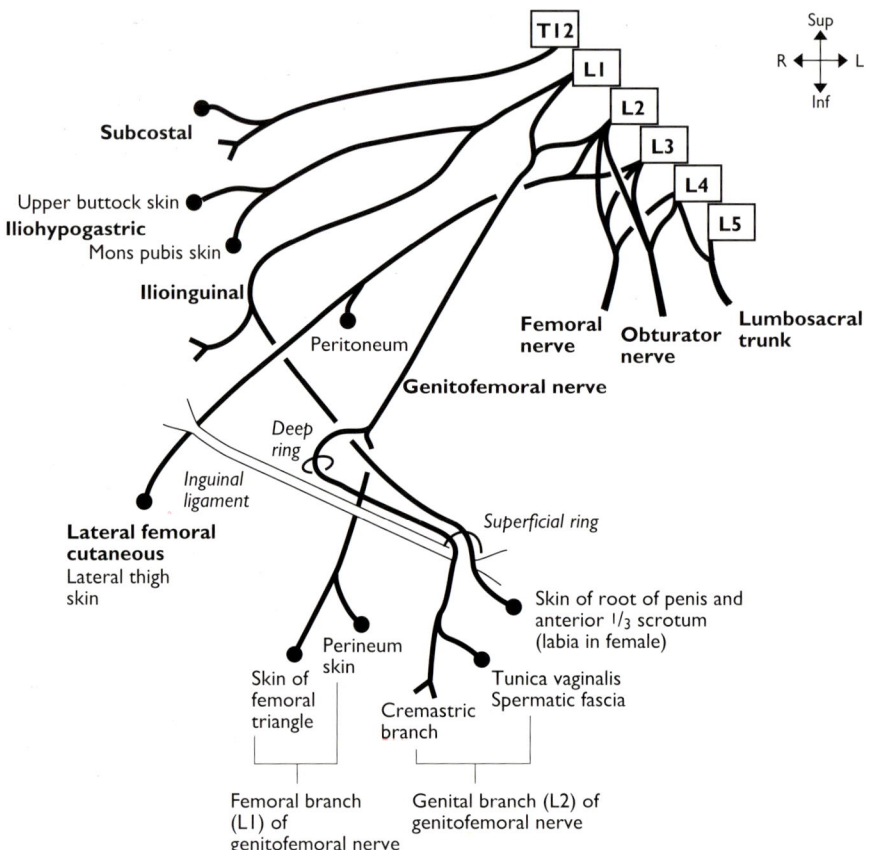

Lumbar plexus (T12,L1,2,3,4,5)

LUMBAR PLEXUS (T12,L1,2,3,4,5)
From: Ant primary rami of T12,L1,2,3,4,5
To: Ns as shown

The plexus is formed within the substance of psoas major anterior to the transverse processes of L2−5 from the anterior primary rami as they emerge from the intervertebral foramina.

Iliohypogastric N (L1). Emerges lateral to psoas on the lumbar fascia at the level of L2 to pass posterior to the lower pole of the kidney and over quadratus lumborum. It passes above the iliac crest, between transversus and internal oblique abdominis to pierce the latter above the anterior superior iliac spine. It supplies both muscles before becoming cutaneous.

Ilioinguinal N (L1). Emerges lateral to psoas on the lumbar fascia, passes posterior to the lower pole of the kidney, over quadratus lumborum and penetrates transversus and internal oblique abdominis above the anterior superior iliac spine. It supplies the lowest fibres of these muscles and the conjoint tendon. Its terminal branch enters the inguinal canal from above to pass through the superficial inguinal ring before piercing the external spermatic fascia to become subcutaneous.

Lateral femoral cutaneous N (L2,3). Emerges lateral to psoas below the iliac crest, passes over iliacus obliquely lying posterior to the caecum on the right and descending colon on the left. It runs forward to the anterior superior iliac spine where it penetrates the inguinal ligament at its attachment to pass into the subcutaneous tissue of the lateral thigh.

Genitofemoral N (L1,2). Emerges onto the anteromedial surface of psoas lying posterior to the ureter, gonadal and iliocolic vessels on the right and ureter, gonadal and lower left colic vessels on the left. It divides into genital and femoral branches on the anterior aspect of psoas.

Genital branch. Crosses the external iliac artery, passes through the deep inguinal ring into the inguinal canal and through the superficial inguinal ring to terminate in the spermatic cord and scrotal skin in the male and the skin of mons and labium majus in the female.

Femoral branch. Continues down lateral to the external iliac artery, under the inguinal ligament and into the femoral sheath which it penetrates anteriorly to become subcutaneous.

Lumbosacral trunk (L4,5). Emerges deep from the medial aspect of psoas to pass over the pelvic brim to form the upper fibres of the sciatic N.

6

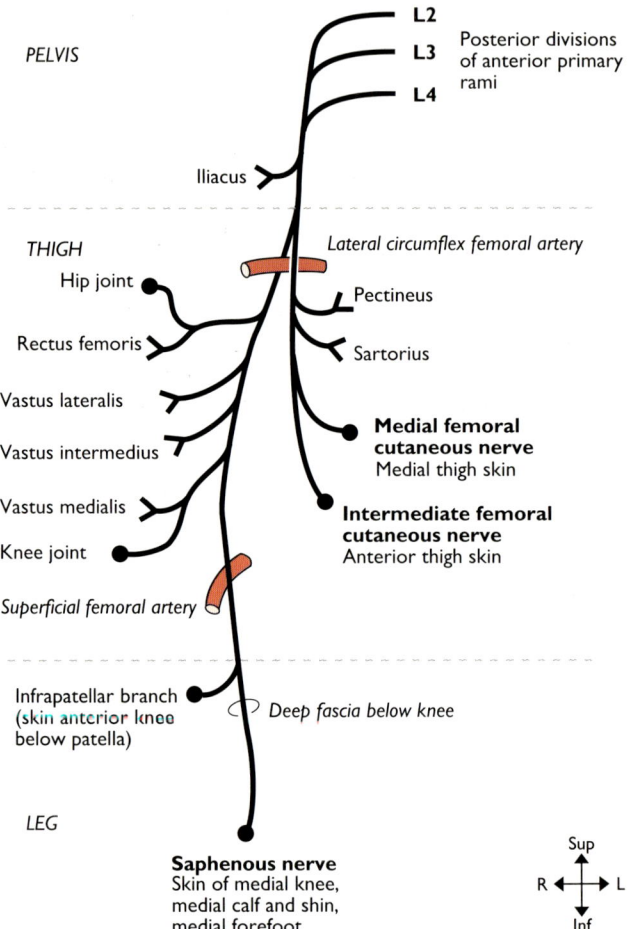

Femoral nerve (L2,3,4)

FEMORAL NERVE (L2,3,4)
From: Post div of ant primary rami of L2,3,4
To: Terminal brs

It is formed within psoas major and emerges from its lateral border low down in the iliac fossa to lie in the groove between psoas and iliacus. It reaches the thigh beneath the inguinal ligament lateral to the femoral artery lying on the tendon of iliacus and psoas. In the femoral triangle it splits into anterior and posterior divisions which straddle the lateral circumflex femoral artery. There are usually four short superficial branches. The deep branches continue down the femoral triangle, the N to vastus medialis running lateral to the femoral artery as far as the upper part of the adductor (Hunter's) canal before entering the muscle.

Saphenous N (post division). Descends in the femoral triangle to reach the adductor canal where it spirals over the femoral artery to lie medial to it. It pierces the deep fascia through the apex of the canal and runs posterior to sartorius to continue with the long saphenous vein. It passes over the subcutaneous surface of the tibia where it is palpable and continues anterior to the medial malleolus closely related to the long saphenous vein. It terminates in branches over the dorsum of the foot.

6

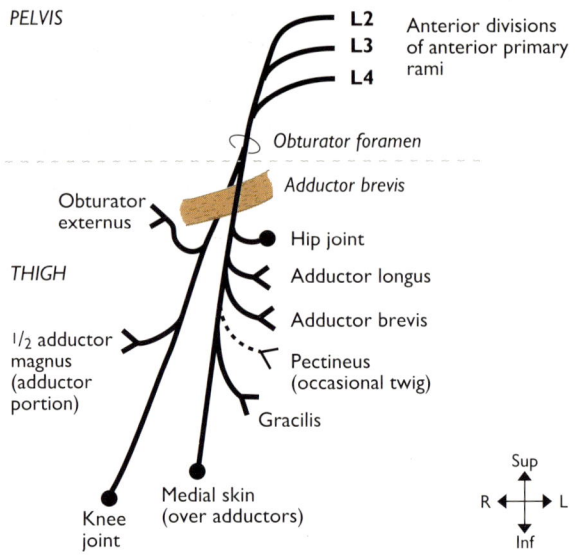

PELVIS

L2
L3
L4

Anterior divisions
of anterior primary
rami

Obturator foramen

Adductor brevis

Obturator
externus

Hip joint

THIGH

Adductor longus

Adductor brevis

¹/₂ adductor
magnus
(adductor
portion)

Pectineus
(occasional twig)

Gracilis

Sup

R ←——→ L

Knee
joint

Medial skin
(over adductors)

Inf

Obturator nerve (L2,3,4)

6

OBTURATOR NERVE (L2,3,4)
From: Ant div of ant primary rami of L2,3,4
To: Terminal brs

This nerve is formed within psoas major and emerges from the medial aspect of the muscle on the ala of the sacrum to pass behind the common iliac vessels. It runs over the pelvic brim on the lateral wall of the pelvis and over the upper fibres of obturator internus to pass through the upper anterior aspect of the obturator foramen. It divides into anterior and posterior divisions which straddle adductor brevis. The posterior division pierces a few fibres of obturator externus and runs deep to adductor brevis on adductor magnus. The anterior division runs on the anterior aspect of adductor brevis deep to pectineus and then deep to adductor longus to end by becoming subcutaneous at the lower border of adductor longus.

6

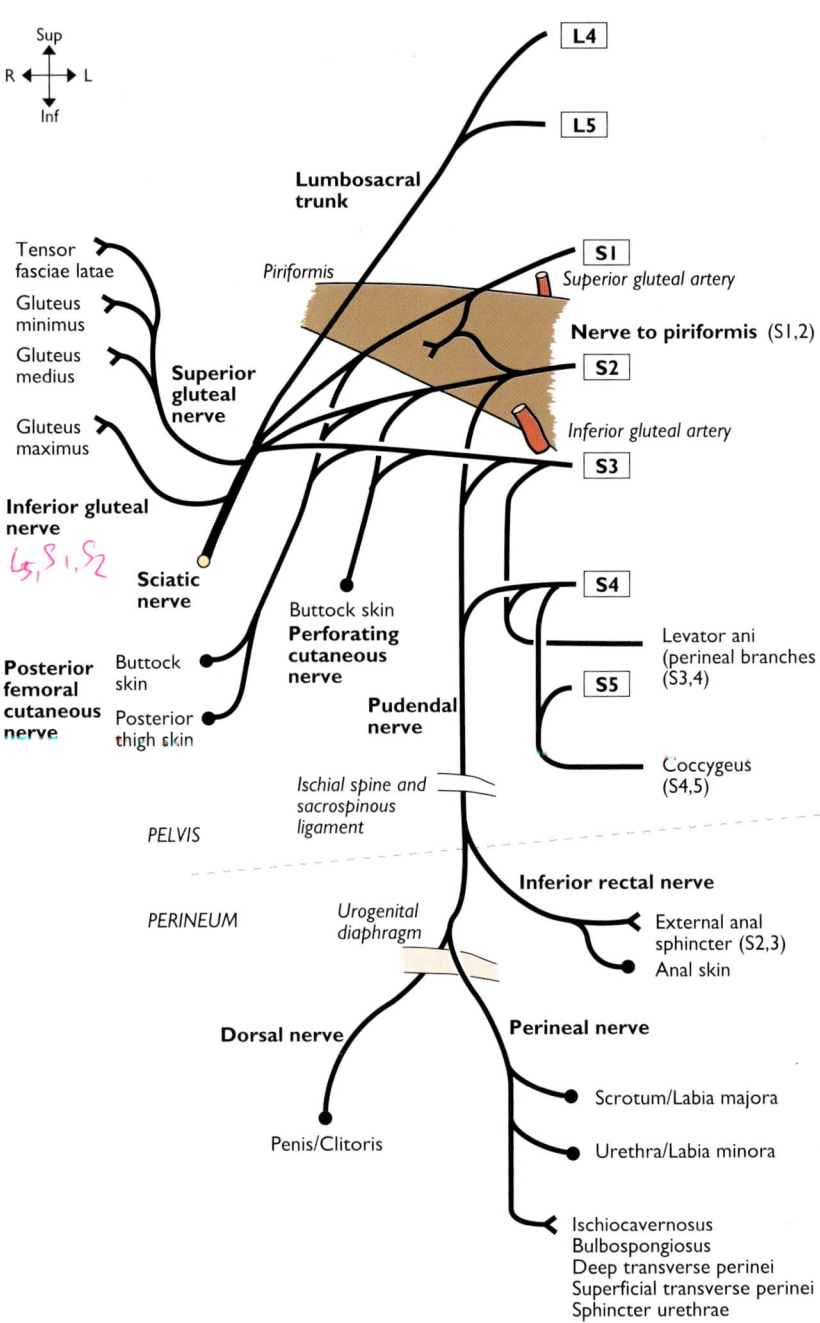

Sup

R ← → L

Inf

L4

L5

Lumbosacral trunk

Piriformis

S1

Superior gluteal artery

Tensor fasciae latae

Gluteus minimus

Nerve to piriformis (S1,2)

Gluteus medius

Superior gluteal nerve

S2

Inferior gluteal artery

Gluteus maximus

S3

Inferior gluteal nerve

L5, S1, S2

Sciatic nerve

Buttock skin
Perforating cutaneous nerve

S4

Levator ani (perineal branches (S3,4)

Posterior femoral cutaneous nerve

Buttock skin

Posterior thigh skin

S5

Pudendal nerve

Ischial spine and sacrospinous ligament

Coccygeus (S4,5)

PELVIS

Inferior rectal nerve

PERINEUM

Urogenital diaphragm

External anal sphincter (S2,3)

Anal skin

Dorsal nerve

Perineal nerve

Scrotum/Labia majora

Urethra/Labia minora

Penis/Clitoris

Ischiocavernosus
Bulbospongiosus
Deep transverse perinei
Superficial transverse perinei
Sphincter urethrae

Sacral plexus (L4,5,S1,2,3,4,5)

SACRAL PLEXUS (L4,5,S1,2,3,4,5)
From: Lumbosacral trunk (L4,5) & ant
  primary rami from L4,5,S1,2,3,4,5
To: Definitive Ns

Lies on piriformis on the posterior wall of
the pelvis deep to the internal iliac vessels
(and the sigmoid vessels on the left) and is
protected by a sheet of pelvic fascia overlying it. Its roots are characteristically related to arteries which pass between them as
shown.

Superior gluteal N (L4,5,S1). Emerges from
the upper roots of the sciatic N and passes
out of the pelvis above piriformis through
the greater sciatic foramen. It runs between
gluteus medius and minimus over the middle
gluteal line on the outer surface of the ilium
to terminate in muscular branches.

Inferior gluteal N (L5,S1,2). Emerges from
the middle roots of the sciatic N and passes
out of the pelvis below piriformis through
the greater sciatic foramen to enter gluteus
maximus.

Posterior femoral cutaneous N (S1,2,3).
Passes out of the pelvis below piriformis
through the greater sciatic foramen. It runs
on the sciatic N, over the long head of biceps
femoris to become subcutaneous extending
as far as the popliteal fossa.

Perforating cutaneous N (S2,3). Passes
through the sacrotuberous ligament and
gluteus maximus to become subcutaneous in
the buttock.

Pudendal N (S2,3,4). Passes out of the pelvis
over the sacrospinous ligament close to the
ischial spine through the greater, and re-
entering through the lesser, sciatic foramena.
It runs on the medial surface of the lower
fibres of obturator internus in the pudendal
(Alcock's) canal. It passes in the lateral wall
of the ischio-anal fossa where it gives off its
inferior rectal branch. It passes into the
perineum and gives its terminal branches,
the perineal N being superficial to the
urogenital. diaphragm and the dorsal N deep
to it.

6

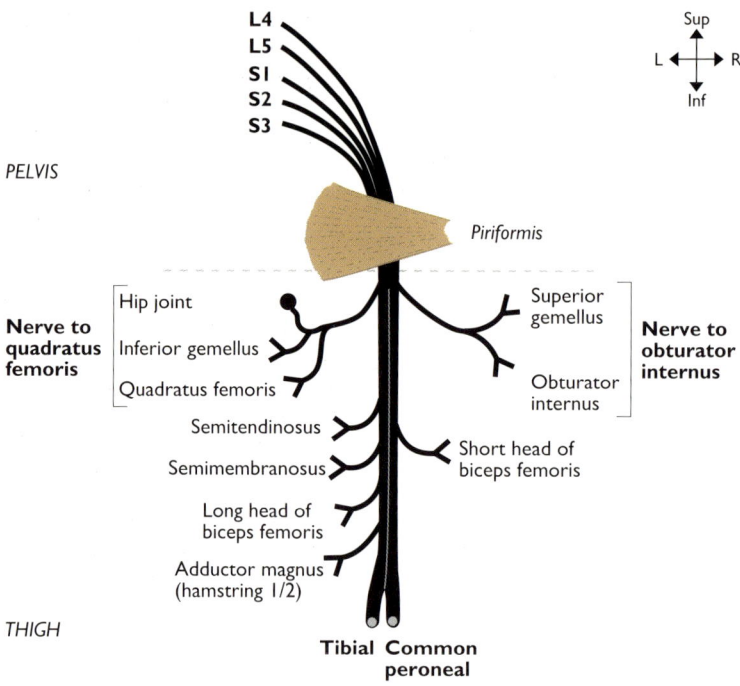

Sciatic nerve (L4,5,S1,2,3)
Note: viewed from behind

SCIATIC NERVE (L4,5,S1,2,3)
From: Ant primary rami of L4,5,S1,2,3
To: Tibial & common peroneal Ns

It is formed in the upper sacral plexus and passes out of the greater sciatic foramen below piriformis. In the buttock and thigh it lies initially deep to gluteus maximus lying on gemellus superior, obturator internus tendon and gemellus inferior and then on quadratus femoris and adductor magnus. It passes out of the cover of gluteus maximus and for a short distance it is covered by only deep fascia, before it passes deep to the two heads of biceps femoris. It runs vertically down in the midline of the posterior compartment of the thigh and terminates by dividing into common peroneal and tibial Ns usually two thirds of the way down the thigh. In its course over the gemelli it is a close posterior relation of the ischium and posterior rim of the acetabulum.

N to quadratus femoris (L4,5,S1). Arises from the anterior surface of the sciatic N in the pelvis and leaves the pelvis in this position through the greater sciatic foramen, lying between the sciatic N and the ischium. Running deep to the tendon of obturator internus and the gemelli it supplies gemellus inferior before passing into quadratus femoris from above.

N to obturator internus (L5,S1,2). Arises from the anterior surface of the sciatic N in the pelvis and passes medially to leave the pelvis through the greater sciatic foramen below piriformis and medial to the sciatic N. It passes over the ischial spine (lateral to the pudendal neurovascular bundle) and sends a branch to gemellus superior before turning forward to pass through the lesser sciatic foramen, penetrating and supplying obturator internus as it does so.

6

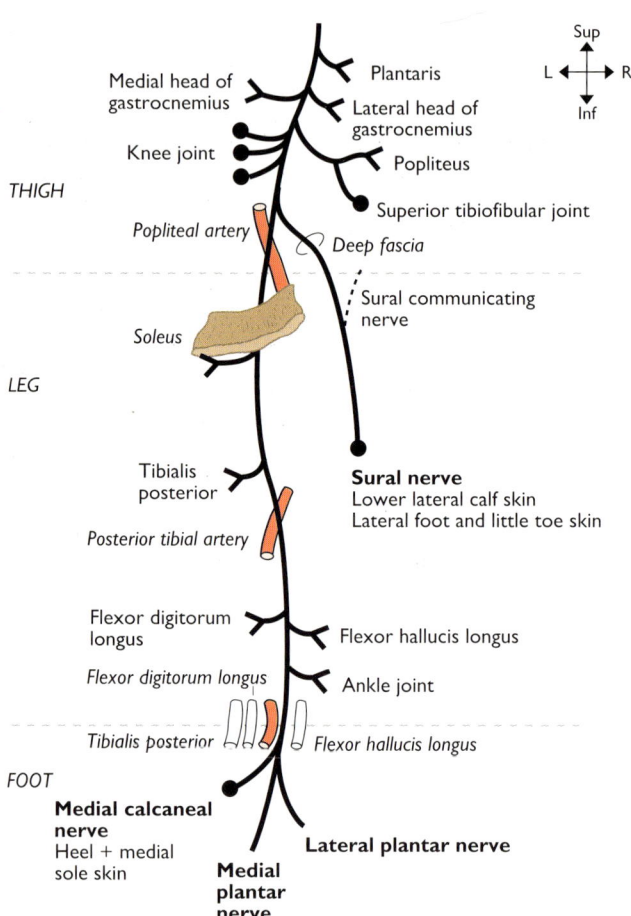

Medial head of
gastrocnemius

Plantaris

Lateral head of
gastrocnemius

Knee joint

Popliteus

THIGH

Popliteal artery

Superior tibiofibular joint

Deep fascia

Soleus

Sural communicating
nerve

LEG

Tibialis
posterior

Sural nerve
Lower lateral calf skin
Lateral foot and little toe skin

Posterior tibial artery

Flexor digitorum
longus

Flexor hallucis longus

Flexor digitorum longus

Ankle joint

Tibialis posterior

Flexor hallucis longus

FOOT

**Medial calcaneal
nerve**
Heel + medial
sole skin

Lateral plantar nerve

**Medial
plantar
nerve**

Sup

L ← → R

Inf

Tibial nerve (L4,5,S1,2,3)
Note: viewed from behind

6

TIBIAL NERVE (L4,5,S1,2,3)
From: Sciatic N
To: Med & lat plantar Ns

It arises in the lower third of the thigh above the apex of the popliteal fossa as the larger terminal branch of the sciatic N, and passes down in the midline into the fossa between semitendinosus and biceps femoris, lying deep to them. It lies markedly lateral to the popliteal artery on entry to the fossa but then the artery crosses deep to the N to lie lateral to it. The tibial N and the popliteal artery remain separated by the popliteal veins. The nerve leaves the fossa deep to the two heads of gastrocnemius by passing deep to the fibrous arch of soleus. It then runs deep to soleus on tibialis posterior in the midline, crossing over the posterior tibial artery from medial to lateral half way down the calf. It slopes gently medially in the lower calf passing behind the medial malleolus of the lower tibia between the posterior tibial artery medially and the tendon of flexor hallucis longus laterally. It runs under the flexor retinaculum, over the medial tubercle of the posterior process of the talus and divides into terminal branches.

Sural N. Arises in the popliteal fossa, passing out over the 'V' behind the two heads of gastrocnemius and is joined by the sural communicating N from the common peroneal N. It pierces the deep fascia to become subcutaneous. It runs down laterally accompanied by the short saphenous vein to pass behind the lateral malleolus over the superior peroneal retinaculum to end in the lateral skin over the foot as terminal branches.

6

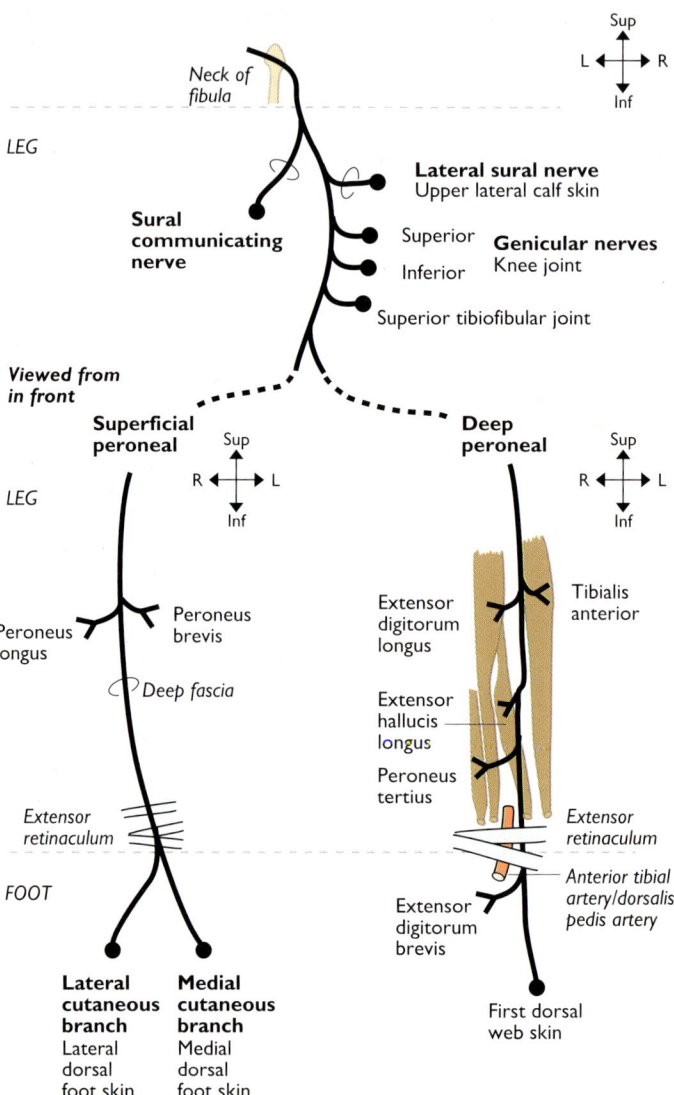

Common, superficial and deep peroneal nerves

Note: the common peroneal nerve is viewed from behind but the superficial deep nerves are viewed from in front

COMMON PERONEAL NERVE
(L4,5,S1,2)
From: Sciatic N
To: Superficial & deep peroneal Ns

It arises in the lower third of the thigh above the apex of the popliteal fossa as the smaller terminal branch of the sciatic N. It passes into the popliteal fossa along the upper lateral boundary just beneath the edge of biceps femoris and runs over plantaris, the lateral head of gastrocnemius and the posterior capsule of the knee joint. It runs over the fibular attachment of soleus to wind around the neck of the fibula from posterior to lateral. It passes deep to peroneus longus where it divides.

SUPERFICIAL PERONEAL NERVE
(L5,S1,2)
From: Common peroneal N
To: Terminal brs

It arises deep to peroneus longus and passes forwards and downwards to lie over the lateral surface of the fibula between peroneus longus and brevis. It pierces the deep fascia half way down the leg to become subcutaneous. It runs downwards superficial to the superior and inferior extensor retinacula to end as terminal branches over them.

DEEP PERONEAL NERVE
(L4,5,S1,2)
From: Common peroneal N
To: Terminal brs

It arises deep to peroneus longus and passes forwards deep to the muscle to wind around the fibula and to pass through the anterior intermuscular septum. It continues deep to extensor digitorum longus to appear between it and tibialis anterior lying on the interosseous membrane in the upper quarter of the anterior compartment. It runs down the interosseous membrane with the anterior tibial vessels, coming to lie between extensor hallucis longus and tibialis anterior in the lower three quarters of the compartment. It passes anterior to the tibia at the ankle joint between the anterior tibial artery medially and the tendon of extensor hallucis longus laterally, running beneath the superior and inferior extensor retinacula. It breaks up into terminal branches on the dorsum of the foot. (Articular branch to ankle joint not shown.)

6

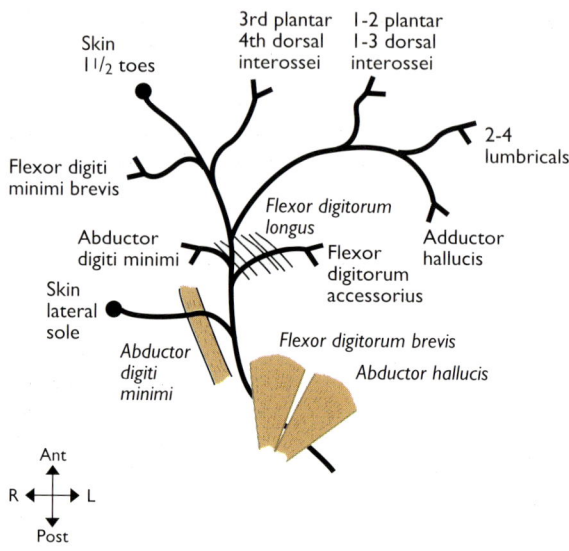

Lateral plantar nerve (S1,2)

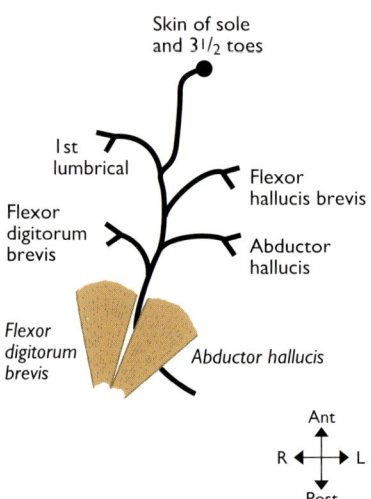

Medial plantar nerve (L4,5)

6

LATERAL PLANTAR NERVE (S1,2)
From: Tibial N
To: Terminal brs

It arises beneath the flexor retinaculum and runs forward with the lateral plantar artery around the sustentaculum tali of the calcaneus deep to abductor hallucis. It runs over the origin of flexor digitorum accessorius beneath flexor digitorum brevis, and its superficial terminal branches appear more superficially between flexor digitorum brevis and abductor digiti minimi. Its deep terminal branches run medially beneath the long flexor tendons and across the metatarsal shafts to end in muscular branches.

MEDIAL PLANTAR NERVE (L4,5)
From: Tibial N
To: Terminal brs

It arises beneath the flexor retinaculum and runs with the medial plantar artery around the sustentaculum tali of the calcaneus deep to abductor hallucis. It pierces the plantar fascia in so doing and runs forward over the tendon of flexor digitorum longus to appear more superficially again between abductor hallucis and flexor digitorum brevis in the sole of the foot.

6

6

7: DERMATOMES AND CUTANEOUS NERVE DISTRIBUTION

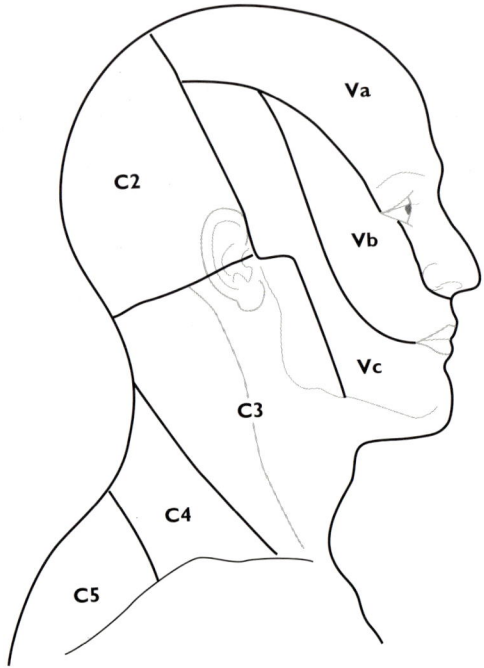

Dermatomes: head and neck

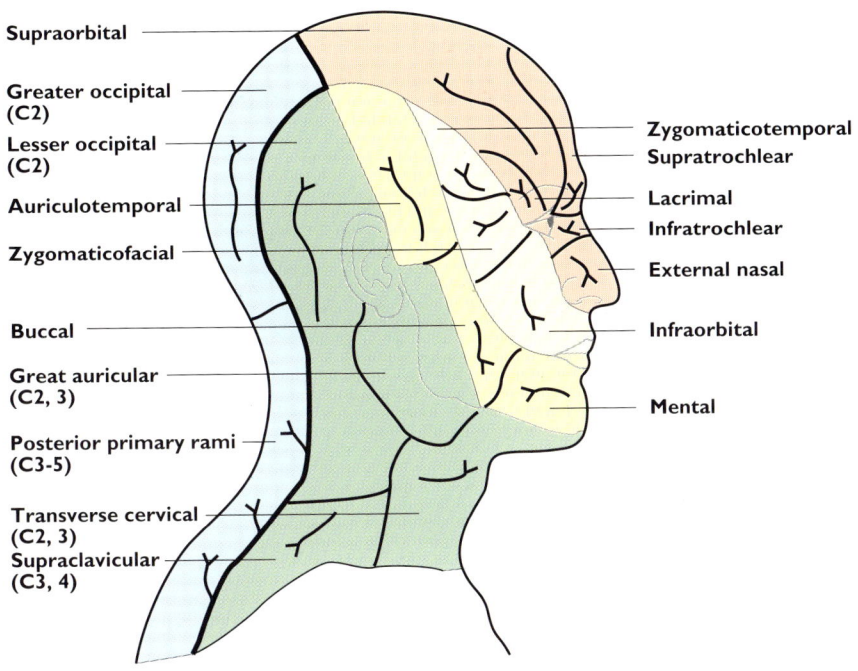

Supraorbital

Greater occipital
(C2)

Lesser occipital
(C2)

Auriculotemporal

Zygomaticofacial

Buccal

Great auricular
(C2, 3)

Posterior primary rami
(C3-5)

Transverse cervical
(C2, 3)

Supraclavicular
(C3, 4)

Zygomaticotemporal
Supratrochlear

Lacrimal

Infratrochlear

External nasal

Infraorbital

Mental

Cutaneous nerves: head and neck

7

7

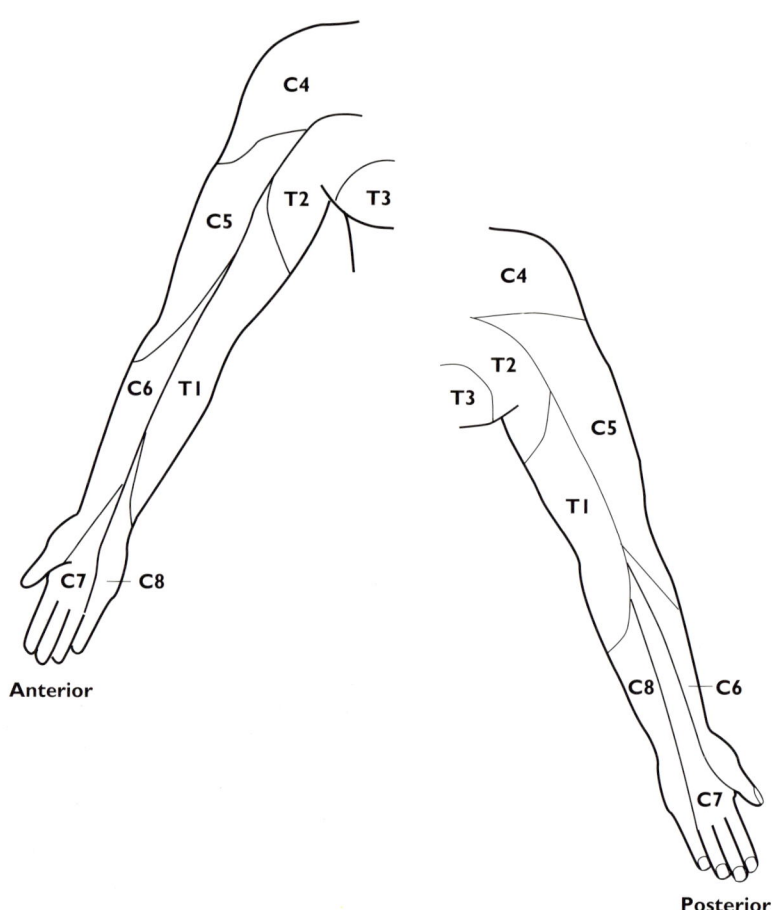

Anterior

Posterior

Dermatomes: arm

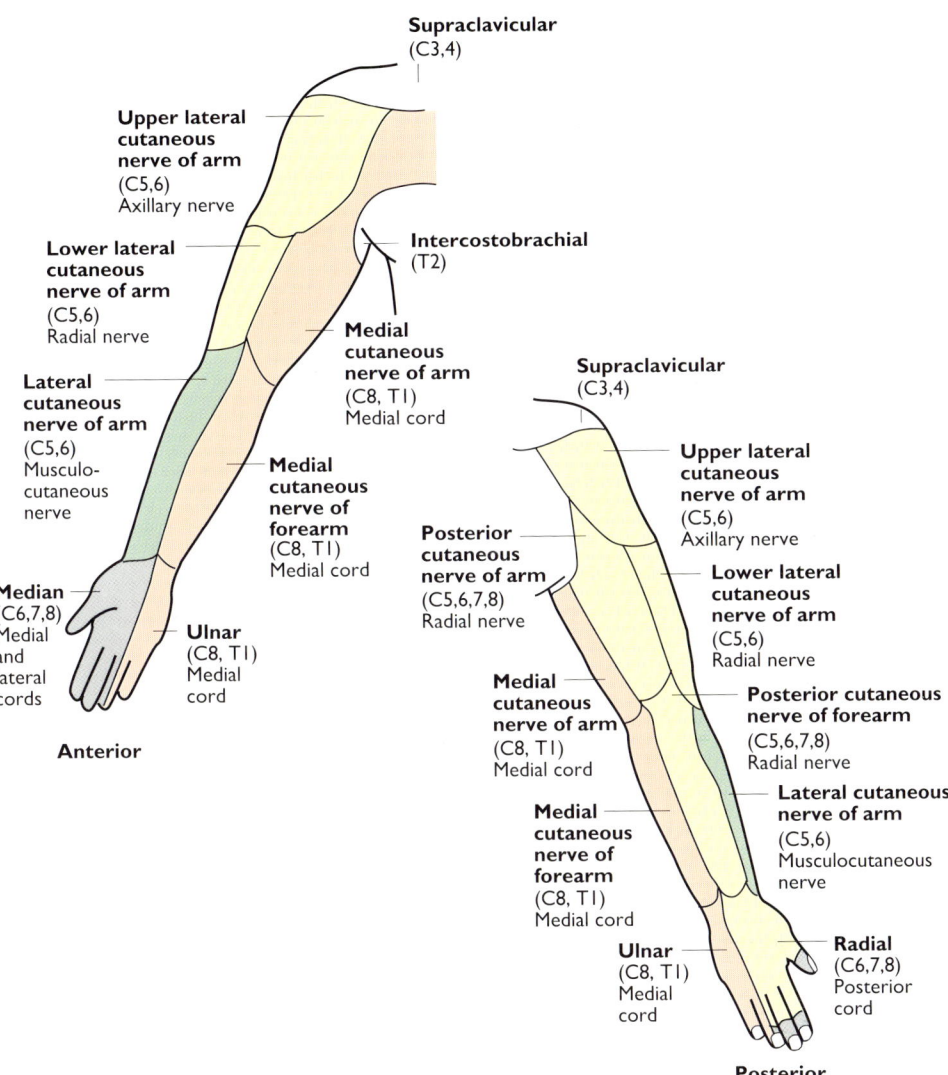

Supraclavicular
(C3,4)

Upper lateral cutaneous nerve of arm
(C5,6)
Axillary nerve

Intercostobrachial
(T2)

Lower lateral cutaneous nerve of arm
(C5,6)
Radial nerve

Medial cutaneous nerve of arm
(C8, T1)
Medial cord

Supraclavicular
(C3,4)

Lateral cutaneous nerve of arm
(C5,6)
Musculo-cutaneous nerve

Medial cutaneous nerve of forearm
(C8, T1)
Medial cord

Upper lateral cutaneous nerve of arm
(C5,6)
Axillary nerve

Posterior cutaneous nerve of arm
(C5,6,7,8)
Radial nerve

Lower lateral cutaneous nerve of arm
(C5,6)
Radial nerve

Median
(C6,7,8)
Medial and lateral cords

Ulnar
(C8, T1)
Medial cord

Medial cutaneous nerve of arm
(C8, T1)
Medial cord

Posterior cutaneous nerve of forearm
(C5,6,7,8)
Radial nerve

Anterior

Lateral cutaneous nerve of arm
(C5,6)
Musculocutaneous nerve

Medial cutaneous nerve of forearm
(C8, T1)
Medial cord

Ulnar
(C8, T1)
Medial cord

Radial
(C6,7,8)
Posterior cord

Posterior

Cutaneous nerves: arm

7

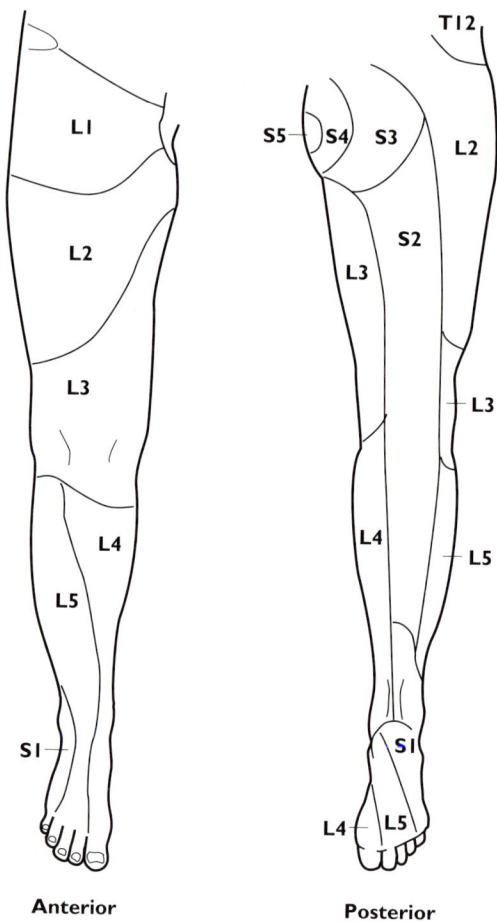

Anterior

Posterior

Dermatomes: leg

7

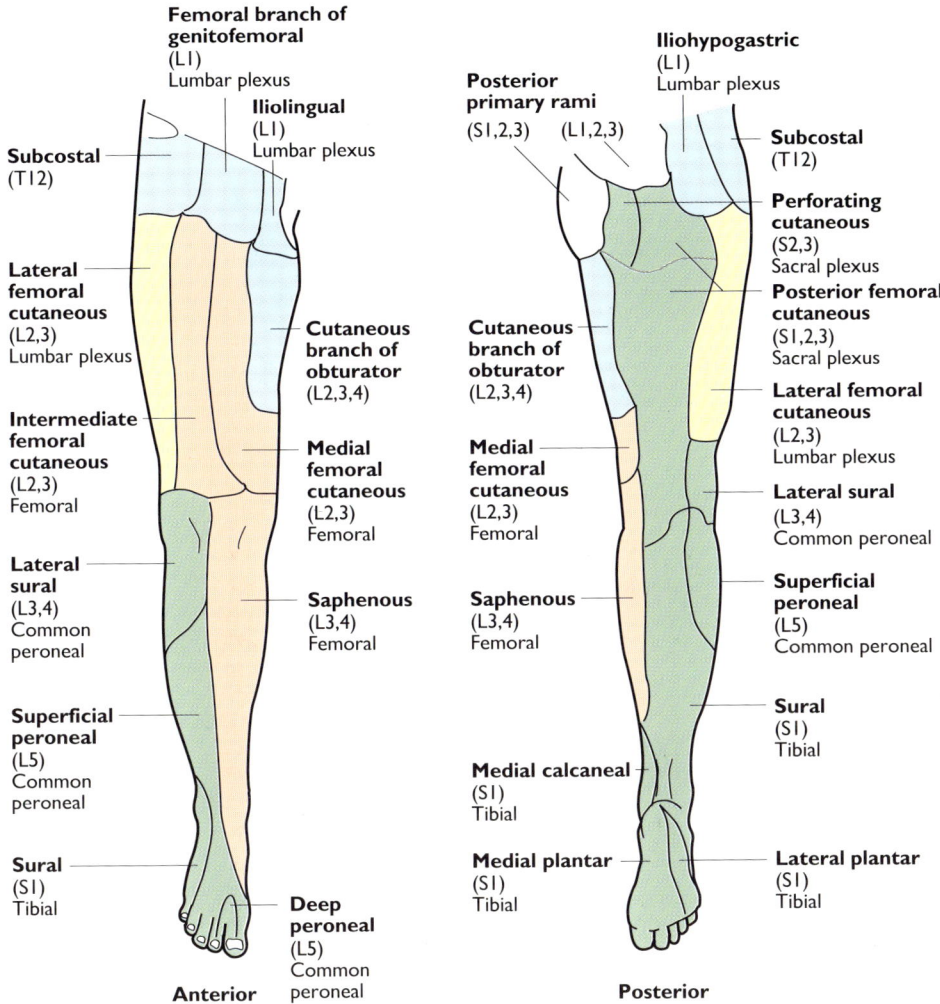

Cutaneous nerves: leg

7

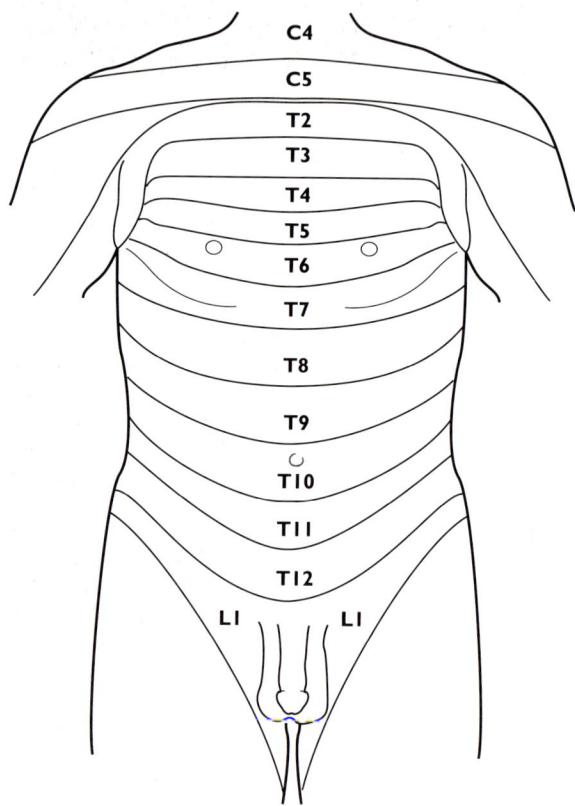

Dermatomes: thorax and abdomen

8

8

ABDUCTOR DIGITI MINIMI (foot)

Arises Med & lat processes of post calcaneal tuberosity

Inserts Lat side of base of prox phalanx of 5th toe & 5th MT

Action Flexes & abducts 5th toe. Supports lat longitudinal arch

Nerve Lat plantar N (S2,3)

ABDUCTOR DIGITI MINIMI (hand)

Arises Pisiform bone, pisohamate lig & flexor retinaculum

Inserts Ulnar side of base of prox phalanx of little finger & extensor expansion (±sesamoid)

Action Abducts little finger at MCP jnt

Nerve Deep br of ulnar N (C8,T1)

ABDUCTOR HALLUCIS

Arises Med process of post calcaneal tuberosity & flexor retinaculum

Inserts Med aspect of base of prox phalanx of big toe via med sesamoid

Action Flexes & abducts big toe. Supports med longitudinal arch

Nerve Med plantar N (S1,2)

ABDUCTOR POLLICIS BREVIS

Arises Tubercle of scaphoid & flexor retinaculum

Inserts Radial sesamoid of prox phalanx of thumb & tendon of extensor pollicis longus

Action Abducts thumb at MCP & CMC jnts

Nerve Recurrent (muscular) br of median N (C8,T1) occasionally by ulnar N

ABDUCTOR POLLICIS LONGUS

Arises Upper post surface of ulna & middle third of post surface of radius & interosseous membrane between

Inserts Over tendons of radial extensors & brachioradialis to base of 1st MC & trapezium

Action Abducts & extends thumb at CMC jnt

Nerve Post interosseous N (C6,7)

Notes Forms radial side of snuff box

ADDUCTOR BREVIS

Arises Inf ramus & body of pubis

Inserts Upper third of linea aspera

Action Adducts hip

Nerve Ant div of obturator N (L2,3)

ADDUCTOR HALLUCIS

Arises Oblique head: base of 2, 3, 4 MTs. Transverse head: plantar MT ligs & deep transverse lig

Inserts Lat side of base of prox phalanx of big toe & lat sesamoid

Action Adducts & flexes MTP jnt of big toe. Supports transverse arch

Nerve Deep br of lat plantar N (S2,3)

Notes If muscle fibres are attached to 1st MT it can be regarded as opponens hallucis

ADDUCTOR LONGUS

Arises Body of pubis inf & med to pubic tubercle

Inserts Lower two thirds of med linea aspera

Action Adducts & med rotates hip

Nerve Ant div of obturator N (L2,3)

ADDUCTOR MAGNUS

Arises Adductor portion: ischiopubic ramus. Hamstring portion: lower outer quadrant of post surface of ischial tuberosity

Inserts Adductor portion: lower gluteal line & linea aspera. Hamstring portion: adductor tubercle

Action Adductor portion: adducts & med rotates hip. Hamstring portion: extends hip

Nerve Adductor portion: post div of obturator N (L2−4). Hamstring portion: tibial portion of sciatic (L4−S3)

8

ADDUCTOR POLLICIS
Arises Oblique head: base of 2nd & 3rd MCs, trapezoid & capitate. Transverse head: palmar border & shaft of 3rd MC
Inserts Ulnar sesamoid then ulnar side of base of prox phalanx & tendon of extensor pollicis longus
Action Adducts CMC jnt of thumb
Nerve Deep br of ulnar N (T1)

ANCONEUS
Arises Smooth surface at lower extremity of post aspect of lat epicondyle of humerus
Inserts Lat side of olecranon
Action Weak extensor of elbow. Moves (abducts) ulna in pronation
Nerve Radial N (C7,8)

ARTICULARIS CUBITI (subanconeus)
Arises Deep distal surface of med head of triceps
Inserts Post capsule of elbow joint
Action Lifts capsule away from joint
Nerve Radial (C6,7,8)

ARTICULARIS GENU
Arises Two slips from ant femur below vastus intermedius
Inserts Apex of suprapatellar bursa
Action Retracts bursa as knee extends
Nerve Post div of femoral N (L3,4)

ARYEPIGLOTTICUS
Arises Apex of arytenoid cartilage
Inserts Lat border of epiglottis
Action Aids closure of additus of larynx
Nerve Recurrent laryngeal br of vagus N (X)

AURICULARIS
Arises Cartilage of auricle
Inserts Galeal aponeurosis
Action Adjusts position of ear
Nerve Temporal br of facial N (VII)

BICEPS BRACHII
Arises Long head: supraglenoid tubercle of scapula. Short head: coracoid process of scapula with coracobrachialis
Inserts Post border of bicipital tuberosity of radius (over bursa) & bicipital aponeurosis to deep fascia & subcutaneous ulna
Action Supinates forearm, flexes elbow, weakly flexes shoulder
Nerve Musculocutaneous N (C5,6) (from lat cord)
Notes Supinates most effectively when elbow flexed

BICEPS FEMORIS
Arises Long head: upper inner quadrant of post surface of ischial tuberosity. Short head: middle third of linea aspera, lat supracondylar ridge of femur
Inserts Styloid process of head of fibula, lat collateral lig & lat tibial condyle
Action Flexes & lat rotates knee. Long head extends hip
Nerve Long head: tibial portion of sciatic N. Short head: common peroneal portion of sciatic N (both L5,S1)

BRACHIALIS
Arises Ant lower half of humerus & med & lat intermuscular septa
Inserts Coranoid process & tuberosity of ulna
Action Flexes elbow
Nerve Musculocutaneous N (C5,6) (from lat cord). Also small supply from radial N (C7)

BRACHIORADIALIS
Arises Upper two thirds of lat supracondylar ridge of humerus & lat intermuscular septum
Inserts Base of styloid process of radius
Action Flexes arm at elbow & brings forearm into midprone position
Nerve Radial N (C5,6)
Notes Overlies radial N & art as they lie on supinator

8

BUCCINATOR
Arises Ext alveolar margins of maxilla & mandible by molar teeth, to maxillary tubercle & pterygoid hamulus & post mylohyoid line respectively, then via pterygomandibular raphe between bones
Inserts Decussates at modiolus of mouth & interdigitates with opposite side
Action Aids mastication, tenses cheeks in blowing & whistling, aids closure of mouth
Nerve Buccal br of facial N (VII)

BULBOSPONGIOSUS
Arises Perineal body (& midline raphe over corpus spongiosum in male)
Inserts Superficial perineal membrane & dorsal penile/clitoral aponeurosis
Action Male: aids emptying of urine & ejaculate from urethra. Female: closes vaginal introitus
Nerve Perineal br of pudendal N (S2,3,4)

CONSTRICTOR OF PHARYNX – INFERIOR
Arises Cricopharyngeus: lat aspect of arch of cricoid cartilage. Thyropharyngeus: oblique line of laminar of thyroid cartilage & fibrous cricothyroid arch
Inserts Cricopharyngeus: continuous with muscle of opposite side, behind pharynx. Thyropharyngeus: lower pharyngeal raphe
Action Aids swallowing. Cricopharyngeus acts as upper oesophageal sphincter
Nerve Pharyngeal plexus (IX, X & sympathetic) via pharyngeal br of X with its motor fibres from cranial accessory (XI)
Notes Killian's dehiscence is between the two parts post

CONSTRICTOR OF PHARYNX – MIDDLE
Arises Lower third of stylohyoid lig, lesser cornu & sup border of greater cornu of hyoid bone
Inserts Middle portion of pharyngeal raphe

Action Aids swallowing
Nerve Pharyngeal plexus (IX, X & sympathetic) via pharyngeal br of X with its motor fibres from cranial accessory N (XI)

CONSTRICTOR OF PHARYNX – SUPERIOR
Arises Lower two thirds of med pterygoid plate, pterygomandibular raphe & post end of mylohyoid line on mandible
Inserts Upper midline pharyngeal raphe & pharyngeal tubercle of clivus of occiput
Action Aids swallowing
Nerve Pharyngeal plexus (IX, X & sympathetic) via pharyngeal br of X with its motor fibres from cranial accessory (XI)

CORACOBRACHIALIS
Arises Coracoid process of scapula with biceps brachii
Inserts Upper half of med border of humerus
Action Flexes & weakly adducts arm
Nerve Musculocutaneous N (C5,6,7) (from lat cord)
Notes Ligament of Struthers as embryological 3rd head. Musculocutaneous N runs through muscle

CORRUGATOR SUPERCILII
Arises Med superciliary arch
Inserts Skin of med forehead
Action Wrinkles forehead
Nerve Temporal br of facial N (VII)

CREMASTER
Arises Lower border of internal oblique & transversus abdominis in inguinal canal
Inserts Loops around spermatic cord & tunica vaginalis & some fibres return to attach to pubic tubercle
Action Retracts testis
Nerve Genital br (L2) of genitofemoral N (L1,2)

8

CRICOTHYROID

Arises Anterolateral aspect of cricoid cartilage

Inserts Inf cornu & lower laminar of thyroid cartilage

Action Lengthens & tenses vocal cords by tilting thyroid cartilage forwards

Nerve Ext br of sup laryngeal br of vagus N (X)

DARTOS

Arises Subcutaneous tissue of scrotum, superficial to superficial fascia (Colles)

Inserts Skin & midline fibrous septum of scrotum

Action Corrugates scrotal skin

Nerve Sympathetic fibres from genital br (L2) of genitofemoral N (L1,2)

DEEP TRANSVERSE PERINEI

Arises Med aspect of ischiopubic ramus & body of ischium

Inserts Midline raphe & perineal body

Action Fixes perineal body & supports pelvic viscera

Nerve Perineal br of pudendal N (S2,3,4)

DELTOID

Arises Lat third of clavicle, acromion, spine of scapula to deltoid tubercle

Inserts Middle of lat surface of humerus (deltoid tuberosity)

Action Abducts arm, ant fibres flex & med rotate, post fibres extend & lat rotate

Nerve Axillary N (C5,6) (from post cord)

DEPRESSOR ANGULI ORIS

Arises Outer surface of mandible post to oblique line

Inserts Modiolus at angle of mouth

Action Depresses & draws angle of mouth laterally

Nerve Mandibular br of facial N (VII)

DEPRESSOR LABII INFERIORIS

Arises Outer surface of mandible along oblique line

Inserts Skin of lower lip

Action Depresses & draws lower lip laterally

Nerve Mandibular br of facial N (VII)

DIAPHRAGM

Arises Vertebral: crura from bodies of L1,2 (left), L1—3 (right). Costal: med & lat arcuate ligs, inner aspect of lower six ribs. Sternal: two slips from post aspect of xiphoid

Inserts Trefoil central tendon

Action Inspiration & assists in raising intra-abdominal pressure

Nerve Phrenic N (motor) (C3,4,5). Sensory: phrenic, intercostals (6—12) & upper two lumbar N roots

DIGASTRIC

Arises Ant belly: digastric fossa on post surface of symphysis menti. Post belly: base of med aspect of mastoid process

Inserts Fibrous loop to lesser cornu of hyoid bone

Action Elevates hyoid bone. Aids swallowing & depresses mandible

Nerve Ant belly: mylohyoid N (Vc). Post belly: facial N (VII)

ERECTOR SPINAE — ILIOCOSTOCERVICALIS

Arises Post angles of ribs

Inserts Transverse processes above & below

Action Extends & lat flexes spine

Nerve Post primary rami

Notes Divided into iliocostalis — lumborum, thoracis & cervicalis

ERECTOR SPINAE — LONGISSIMUS

Arises Transverse processes

Inserts Transverse processes several levels above

Action Extends spine

8

Nerve Post primary rami

Notes Divided into longissimus — thoracis, cervicis & capitis

ERECTOR SPINAE — SPINALIS

Arises Spinous processes

Inserts Spinous processes six levels above

Action Lat flexion of spine

Nerve Post primary rami

Notes Divided into spinalis — thoracis, cervicis & capitis

EXTENSOR CARPI RADIALIS BREVIS

Arises Common extensor origin on ant aspect of lat epicondyle of humerus

Inserts Post base of 3rd MC

Action Extends & abducts hand at wrist

Nerve Post interosseous N (C7,8)

EXTENSOR CARPI RADIALIS LONGUS

Arises Lower third of lat supracondylar ridge of humerus & lat intermuscular septum

Inserts Post base of 2nd MC

Action Extends & abducts hand at wrist

Nerve Radial N (C6,7)

EXTENSOR CARPI ULNARIS

Arises Common extensor origin on ant aspect of lat epicondyle of humerus

Inserts Base of 5th MC via groove by ulnar styloid

Action Extends & adducts hand at wrist

Nerve Post interosseous N (C7,8)

EXTENSOR DIGITI MINIMI (hand)

Arises Common extensor origin on ant aspect of lat epicondyle of humerus

Inserts Extensor expansion of little finger — usually two tendons which are joined by a slip from extensor digitorum at MCP jnt

Action Extends all jnts of little finger

Nerve Post interosseous N (C7,8)

EXTENSOR DIGITORUM (hand)

Arises Common extensor origin on ant aspect of lat epicondyle of humerus

Inserts Ext expansion to middle & distal phalanges by four tendons. Tendons 3 & 4 usually fuse & little finger just receives a slip

Action Extends all jnts of fingers

Nerve Post interosseous N (C7,8)

EXTENSOR DIGITORUM BREVIS (foot)

Arises Sup surface of ant calcaneus

Inserts Four tendons into prox phalanx of big toe & long extensor tendons to toes 2, 3 and 4

Action Extends toes when foot fully dorsiflexed

Nerve Deep peroneal N (L5,S1)

Notes Med one of four tendons could be regarded as extensor hallucis brevis

EXTENSOR DIGITORUM LONGUS (foot)

Arises Upper two thirds of ant shaft of fibula, interosseous membrane & sup tibiofibular jnt

Inserts Extensor expansion of lat four toes

Action Extends toes & extends foot at ankle

Nerve Deep peroneal N (L5,S1)

EXTENSOR HALLUCIS LONGUS

Arises Middle half of ant shaft of fibula

Inserts Base of distal phalanx of great toe

Action Extends big toe & foot. Inverts foot & tightens subtalar jnts

Nerve Deep peroneal N (L5,S1)

EXTENSOR INDICIS

Arises Lower post shaft of ulna (below extensor pollicis longus) & adjacent interosseous membrane

Inserts Extensor expansion of index finger (tendon lies on ulnar side of extensor digitorum tendon)

Action Extends all jnts of index finger

Nerve Post interosseous N (C7,8)

8

EXTENSOR POLLICIS BREVIS

Arises Lower third of post shaft of radius & adjacent interosseous membrane

Inserts Over tendons of radial extensors & brachioradialis to base of prox phalanx of thumb

Action Extends MCP jnt of thumb

Nerve Post interosseous N (C7,8)

Notes Forms radial side of snuff box

EXTENSOR POLLICIS LONGUS

Arises Middle third of post ulna (below abductor pollicis longus) & adjacent interosseous membrane

Inserts Base of distal phalanx of thumb via Lister's tubercle (dorsal tubercle of radius)

Action Extends IP & MCP jnts of thumb

Nerve Post interosseous N (C7,8)

Notes Forms ulnar side of snuff box

EXTERNAL OBLIQUE ABDOMINIS

Arises Ant angles of lower eight ribs

Inserts Outer ant half of iliac crest, inguinal lig, pubic tubercle & crest, & aponeurosis of ant rectus sheath

Action Supports abdominal wall, assists forced expiration, aids raising intra-abdominal pressure &, with muscles of opposite side, abducts & rotates trunk

Nerve Ant primary rami (T7−12)

Notes Interdigitates with four slips of serratus ant & four of latissimus dorsi

FLEXOR CARPI RADIALIS

Arises Common flexor origin of med epicondyle of humerus

Inserts Bases of 2nd & 3rd MCs via groove in trapezium & slip to scaphoid

Action Flexes & abducts wrist

Nerve Median N (C6,7) (from med & lat cords)

FLEXOR CARPI ULNARIS

Arises Humeral head: common flexor origin of med epicondyle. Ulnar head: aponeurosis from med olecranon & upper three quarters subcutaneous border of ulna

Inserts Pisiform, hook of hamate, base of 5th MC via pisohamate & pisometacarpal ligs

Action Flexes & adducts wrist. Fixes pisiform during action of hypothenar muscles

Nerve Ulnar N (C6,7) (by communication from lat cord)

Notes Ulnar N passes between two heads

FLEXOR DIGITI MINIMI

BREVIS (foot)

Arises Base of 5th MT & sheath of peroneus longus

Inserts Lat side of base of prox phalanx of little toe

Action Flexes MTP jnt of little toe

Nerve Superficial br of lat plantar N (S2,3)

Notes A few muscle fibres to distal half of plantar surface of 5th MT represent opponens digiti minimi

FLEXOR DIGITI MINIMI

BREVIS (hand)

Arises Flexor retinaculum & hook of hamate

Inserts Ulnar side of base of prox phalanx of little finger

Action Flexes MCP jnt of little finger

Nerve Deep br of ulnar N (C8,T1)

FLEXOR DIGITORUM ACCESSORIUS (QUADRATUS PLANTAE) (foot)

Arises Med & lat sides of calcaneus

Inserts Tendons of flexor digitorum longus

Action Assists flexor digitorum longus to flex lat four toes, especially when ankle is plantar flexed

Nerve Lat plantar N (S2,3)

FLEXOR DIGITORUM

BREVIS (foot)

Arises Med process of post calcaneal tuberosity

Inserts Four tendons to four lat toes to borders of middle phalanx. Tendons of

8

flexor digitorum longus pass through them

Action Flexes lat four toes. Supports med & lat longitudinal arches

Nerve Med plantar N (S1,2)

FLEXOR DIGITORUM LONGUS (foot)

Arises Post shaft of tibia below soleal line & by broad aponeurosis from fibula

Inserts Base of distal phalanges of lat four toes

Action Flexes distal phalanges of lat four toes & foot at ankle. Supports lat longitudinal arch

Nerve Tibial N (S1,2)

Notes Med two tendons receive slips from flexor hallucis longus & all four receive insertion of flexor accessorius & each give lumbricals

FLEXOR DIGITORUM PROFUNDUS (hand)

Arises Med olecranon, upper three quarters of ant & med surface of ulna as far round as subcutaneous border & narrow strip of interosseous membrane

Inserts Distal phalanges of med four fingers. Tendon to index finger separates early

Action Flexes distal IP jnts, then secondarily flexes prox IP & MCP jnts & wrist

Nerve Median N (ant interosseous) (C6,7)/ ulnar N (C7,8)

Notes Nerve supply as above in 60%. In 40% it is a 3 : 1 ratio either way

FLEXOR DIGITORUM SUPERFICIALIS (hand)

Arises Humeral head: common flexor origin of med epicondyle of humerus, med lig of elbow. Ulnar head: sublime tubercle (med border of coronoid process) & fibrous arch. Radial head: whole length of ant oblique line

Inserts Tendons split to insert onto sides of middle phalanges of med four fingers

Action Flexes prox IP jnts & secondarily MCP jnts & wrist

Nerve Median N (C7,8) (from med & lat cords)

Notes Median N applied to under surface of muscle

FLEXOR HALLUCIS BREVIS

Arises Cuboid, lat cuneiform & tibialis posterior insertion over the two remaining cuneiforms

Inserts Med tendon to med side of base of prox phalanx of big toe. Lat tendon to lat side of same, both via sesamoids

Action Flexes MTP jnt of big toe. Supports med longitudinal arch

Nerve Med plantar N (S2,3)

FLEXOR HALLUCIS LONGUS

Arises Lower two thirds of post fibula between median crest & post border, lower intermuscular septum & apo-neurosis of flexor digitorum longus

Inserts Base of distal phalanx of big toe & slips to mcd two tendons of flexor digitorum longus

Action Flexes distal phalanx of big toe, flexes foot at ankle, supports med logitudinal arch

Nerve Tibial N (S2,3)

FLEXOR POLLICIS BREVIS

Arises Flexor retinaculum & tubercle of trapezium

Inserts Base of prox phalanx of thumb (via radial sesamoid)

Action Flexes MCP jnt of thumb

Nerve Recurrent (muscular) br of median N (C8,T1) (may also be from deep br of ulnar N (T1))

FLEXOR POLLICIS LONGUS

Arises Ant surface of radius below ant oblique line & adjacent interosseous membrane

Inserts Base of distal phalanx of thumb

Action Flexes distal phalanx of thumb

Nerve Ant interosseous N (C7,8)

8

GASTROCNEMIUS
Arises Lat head: post surface of lat condyle of femur & highest of three facets on lat condyle. Med head: post surface of femur above med condyle
Inserts Tendo calcaneus to middle of three facets on post aspect of calcaneus
Action Plantar flexes foot. Flexes knee
Nerve Tibial N (S1,2)
Notes Main propulsive force for jumping

GEMELLUS INFERIOR
Arises Upper border of ischial tuberosity
Inserts Middle part of med aspect of greater trochanter of femur
Action Lat rotates & stabilises hip
Nerve N to quadratus femoris (L4,5,S1)

GEMELLUS SUPERIOR
Arises Spine of ischium
Inserts Middle part of med aspect of greater trochanter of femur
Action Lat rotates & stabilises hip
Nerve N to obturator internus (L5,S1,2)

GENIOGLOSSUS
Arises Sup mental spine on post surface of symphysis menti
Inserts Central mass of tongue & mucous membrane
Action Protracts tongue
Nerve Hypoglossal N (XII)

GENIOHYOID
Arises Inf mental spine on post surface of symphysis menti
Inserts Sup border of body of hyoid bone
Action Elevates & protracts hyoid bone. Depresses mandible
Nerve C1 fibres carried by hypoglossal N (XII)

GLUTEUS MAXIMUS
Arises Outer surface of ilium behind post gluteal line & post third of iliac crest, lumbar fascia, lat mass of sacrum, sacrotuberous lig & coccyx
Inserts Deepest quarter into gluteal tuberosity of femur, remaining three quarters into iliotibial tract (ant surface of lat condyle of tibia)
Action Extends & lat rotates hip. Maintain knee extended via iliotibial tract
Nerve Inf gluteal N (L5,S1,2)
Notes Largest muscle in body

GLUTEUS MEDIUS
Arises Outer surface of ilium between post & middle gluteal lines
Inserts Posterolateral surface of greater trocanter of femur
Action Abducts & med rotates hip. Tilts pelvis on walking
Nerve Sup gluteal N (L4,5,S1)

GLUTEUS MINIMUS
Arises Outer surface of ilium between middle & inf gluteal lines
Inserts Ant surface of greater trochanter of femur
Action Abducts & med rotates hip. Tilts pelvis on walking
Nerve Sup gluteal N (L4,5,S1)

GRACILIS
Arises Outer surface of ischiopubic ramus
Inserts Upper med shaft of tibia below sartorius
Action Adducts hip. Flexes knee & med rotates flexed knee
Nerve Ant div of obturator N (L2,3)

HYOGLOSSUS (& CHONDROGLOSSUS)
Arises Sup border of greater cornu of hyoid bone
Inserts Lat sides of tongue
Action Depresses tongue
Nerve Hypoglossal N (XII)

ILIACUS
Arises Iliac fossa within abdomen
Inserts Lowermost surface of lesser trochanter of femur

8

Action Flexes & med rotates hip

Nerve Femoral N in abdomen (L2,3)

INFERIOR OBLIQUE (see also obliquus capitis inferior)

Arises Orbital surface of maxilla behind orbital margin

Inserts Post/inf quadrant of sclera behind equator of eyeball

Action Elevates eye in adduction. Extorts eye in abduction

Nerve Inf div of oculomotor N (III)

INFERIOR RECTUS

Arises Inf tendinous ring within orbit

Inserts Inf sclera ant to equator of eyeball

Action Depresses eye. Extorts eye in adduction

Nerve Inf div of oculomotor N (III)

INFRASPINATUS

Arises Med three quarters of infraspinous fossa of scapula & fibrous intermuscular septa

Inserts Middle facet of greater tuberosity of humerus & capsule of shoulder jnt

Action Lat rotates arm & stabilises shoulder jnt

Nerve Suprascapular N (C5,6) (from upper trunk)

Notes Bursa under tendon over glenoid angle. Tendon forms part of rotator cuff

INTERCOSTALS EXTERNAL

Arises Inf border of ribs as far forwards as costochondral junctions

Inserts Sup border of ribs below, passing obliquely downwards & forwards

Action Fix intercostal spaces during respiration. Aids forced respiration by elevating ribs

Nerve Muscular collateral brs of intercostal Ns

INTERCOSTALS INNERMOST

Arises Int aspect of ribs above & below

Inserts Int aspect of ribs above & below

Action Fix intercostal spaces during respiration

Nerve Muscular collateral brs of intercostal Ns

INTERCOSTALS INTERNAL

Arises Inf border of ribs as far back as post angles

Inserts Sup border of ribs below, passing obliquely downwards & backwards

Action Fix intercostal spaces during respiration. Aids forced inspiration by elevating ribs

Nerve Muscular collateral brs of intercostal Ns

INTERNAL OBLIQUE ABDOMINIS

Arises Lumbar fascia, ant two thirds of iliac crest & lat two thirds of inguinal lig

Inserts Costal margin, aponeurosis of rectus sheath (ant & post), conjoint tendon to pubic crest & pectineal line

Action Supports abdominal wall, assists forced respiration, aids raising intra-abdominal pressure &, with muscles of other side, abducts & rotates trunk. Conjoint tendon supports post wall of inguinal canal

Nerve Ant primary rami (T7−12) (conjoint tendon ilioinguinal N (L1))

INTEROSSEI − DORSAL OF FOOT (4)

Arises Bipennate from inner aspects of shafts of all MTs

Inserts Bases of prox phalanges & dorsal extensor expansions of med side of 2nd toe & lat sides of 2nd, 3rd & 4th toes

Action Abduct 2nd, 3rd & 4th toes from axis of 2nd toe. Assist lumbricals in extending IP jnts whilst flexing MTP jnts

Nerve Lat plantar N (1−3: deep br; 4: superficial br) (S2,3)

8

INTEROSSEI – DORSAL OF HAND (4)

Arises Bipennate from inner aspects of shafts of all MCs

Inserts Prox phalanges & dorsal extensor expansion on radial side of index & middle fingers & ulnar side of middle & ring fingers

Action Abduct from axis of middle finger. Flex MCP jnt whilst extending IP jnts

Nerve Deep br of ulnar N (T1)

INTEROSSEI – PALMAR OF HAND (4)

Arises Ant shafts of 1, 2, 4, 5 MCs (unipennate)

Inserts Prox phalanges & dorsal extensor expansion on ulnar side of index & radial side of ring & little fingers & to ulnar sesamoid of thumb

Action Adduct to axis of middle finger. Flex MCP jnt whilst extending IP jnts

Nerve Deep br of ulnar N (T1)

Notes 1st palmar interosseous is small and may be absent

INTEROSSEI – PLANTAR OF FOOT (3)

Arises Inferomedial shafts of 3rd, 4th & 5th MTs (single heads)

Inserts Med sides of bases of prox phalanges with slips to dorsal extensor expansions of 3rd, 4th & 5th toes

Action Adduct 3rd, 4th & 5th toes to axis of 2nd toe. Assist lumbricals in extending IP jnts whilst flexing MTP jnts

Nerve Deep br of lat plantar N (S2,3)

INTERSPINALES

Arises Spinous processes

Inserts Spinous processes one above

Action Extension of spine

Nerve Post primary rami

INTERTRANSVERSARII

Arises Transverse processes

Inserts Transverse processes one above

Action Lat flexes spine

Nerve Post primary rami

INTRINSIC MUSCLE OF TONGUE

Arises Sup & inf longitudinal, transverse & vertical elements

Inserts Mucous membrane, septum & other muscles of tongue

Action Alter shape of tongue & so aid mastication, speech & swallowing

Nerve Hypoglossal N (XII)

ISCHIOCAVERNOSUS

Arises Med aspect of ischium & ischiopubic ramus

Inserts Inferolateral aponeurosis over crura of penis/clitoris

Action Stabilises erect penis

Nerve Perineal br of pudendal N (S2,3,4)

LATERAL CRICOARYTENOID

Arises Lat aspect of arch of cricoid cartilage

Inserts Muscular process of arytenoid cartilage .

Action Adducts & med rotates arytenoid cartilage (closes rima glottidis)

Nerve Recurrent laryngeal br of vagus N (X)

LATERAL PTERYGOID

Arises Upper head: infratemporal surface of sphenoid bone. Lower head: lat surface of lat pterygoid plate

Inserts Pterygoid fovea below condyloid process of mandible & intra-articular cartilage of temporomandibular jnt

Action Depresses & protracts mandible to open mouth. Pulls forward cartilage of jnt during opening of mouth

Nerve Ns to lat pterygoid (ant div of mandibular N (Vc))

LATERAL RECTUS

Arises Lat tendinous ring within orbit

Inserts Lat sclera ant to equator of eyeball

8

Action Abducts eye
Nerve Abducent N (VI)

LATISSIMUS DORSI
Arises Spine T7, spinous processes &
 supraspinous ligs of all lower thoracic,
 lumbar & sacral vertebrae, lumbar fascia,
 post third iliac crest, last four ribs
 (interdigitating with ext oblique
 abdominis) & inf angle of scapula
Inserts Floor of bicipital groove of humerus
 after spiraling around teres major
Action Extends, adducts & med rotates
 arm. Costal attachment helps with deep
 inspiration & forced expiration
Nerve Thoracodorsal N (C6,7,8) (from
 post cord)

LEVATOR ANGULI ORIS
Arises Ant surface of maxilla below infra-
 orbital foramen
Inserts Outer end of upper lip & modiolus
Action Elevates angle of mouth
Nerve Buccal br of facial N (VII)

LEVATOR ANI – COCCYGEUS
Arises Sacrospinous lig
Inserts Anococcygeal body & coccyx
Action Supports pelvic viscera
Nerve Ant primary rami (perineal brs) of
 S4,5

LEVATOR ANI – ILIOCOCCYGEUS
Arises Post half of fascial line over
 obturator internus & ischial spine
Inserts Anococcygeal body
Action Supports pelvic viscera
Nerve Ant primary rami (perineal brs) of
 S3,4

LEVATOR ANI – PUBOCOCCYGEUS
Arises Post surface of pubis & ant half of
 fascial line over obturator internus
Inserts Anococcygeal body
Action Supports pelvic viscera
Nerve Ant primary rami (perineal brs) of
 S3,4

LEVATOR ANI – PUBORECTALIS
Arises Post surface of pubis
Inserts Midline sling post to rectum
Action Supports & aids continence of
 rectum by maintaining anorectal angle
Nerve Ant primary rami (perineal brs) of
 S3,4

LEVATOR ANI – PUBOVAGINALIS (LEVATOR PROSTATAE)
Arises Post surface of pubis
Inserts Midline raphe post to vagina/
 prostate
Action Supports ant pelvic viscera
Nerve Ant primary rami (perineal brs) of
 S3,4

LEVATOR LABII SUPERIORIS
Arises Med infra-orbital margin
Inserts Skin & muscle of upper lip
Action Elevates & everts upper lip
Nerve Buccal br of facial N (VII)

LEVATOR LABII SUPERIORIS ALAEQUE NASI
Arises Upper frontal process of maxilla
Inserts Skin of lat nostril & upper lip
Action Dilates nostril & elevates upper lip
Nerve Buccal br of facial N (VII)

LEVATOR PALPEBRAE SUPERIORIS
Arises Inf aspect of lesser wing of sphenoid
 bone
Inserts Sup tarsal plate & skin of upper
 eyelid
Action Elevates & retracts upper eyelid
Nerve Sup div of oculomotor N (III) &
 sympathetic to smooth muscle portion

LEVATOR SCAPULAE
Arises Post tubercles of transverse
 processes of C1–4
Inserts Upper part of med border of
 scapula
Action Raises med border of scapula
Nerve Ant primary rami of C3 & C4 &
 dorsal scapular N (C5)

8

LEVATOR VELI PALATINI
Arises Apex of inf surface of petrous temporal bone & med rim of auditory tube
Inserts Palatine aponeurosis
Action Elevates, retracts & lat deviates soft palate. May open auditory tube on swallowing
Nerve Pharyngeal br of vagus N (X) with its motor fibres from cranial accessory N (XI)

LEVATORES COSTARUM
Arises Transverse processes C7 to T11
Inserts Post surface & angle of rib below
Action Elevates ribs
Nerve Post primary rami

LONGUS CAPITIS
Arises Ant tubercles of transverse processes of C3−6
Inserts Ant basilar occipital bone
Action Flexes of cervical spine & atlanto-occipital jnt
Nerve Ant primary rami of C1−3

LONGUS COLLI
Arises Ant body of T1−3, ant tubercles of transverse processes of C3−7
Inserts Ant arch of atlas (C1) & bodies of C2−4
Action Flexes & rotates cervical spine
Nerve Ant primary rami of C2−6

LUMBRICALS OF FOOT (4)
Arises Lat 3: bipennate origin from cleft between the four tendons of flexor digitorum longus. Med 1: unipennate origin from med aspect of 1st tendon
Inserts Dorsal extensor expansion
Action Extend toes at IP jnts & flex MTP jnts
Nerve First: med plantar N (L4,5). 2−4: deep br of lat plantar N (S2,3)

LUMBRICALS OF HAND (4)
Arises Four tendons of flexor digitorum

profundus. Radial 2: radial side only (unipennate). Ulnar 2: cleft between tendons (bipennate)
Inserts Extensor expansion (dorsum of prox phalanx) of fingers 2−5 radial side
Action Flex MCP jnts & extend IP jnts of fingers
Nerve Lat 2: median N (C8,T1). Med 2: deep br of ulnar N (C8,T1)
Notes 60% have nerve supply as above. 40% have 3 : 1 or 1 : 3

MASSETER
Arises Ant two thirds of zygomatic arch & zygomatic process of maxilla
Inserts Lat surface of angle & lower ramus of mandible
Action Elevates mandible (enables forced closure of mouth)
Nerve Ant div of mandibular N (Vc)

MEDIAL PTERYGOID
Arises Deep head. Med side of lat pterygoid plate & fossa between med & lat plates. Superficial head: Tuberosity of maxilla & pyramidal process of palatine bone
Inserts Med aspect of angle of mandible
Action Elevates, protracts & lat displaces mandible to opposite side for chewing
Nerve N to medial pterygoid (main trunk of mandibular N (Vc))

MEDIAL RECTUS
Arises Med tendinous ring within orbit
Inserts Med sclera ant to equator of eyeball
Action Adducts eye
Nerve Inf div of oculomotor N (III)

MENTALIS
Arises Incisive fossa on ant aspect of mandible
Inserts Skin of chin
Action Elevates & wrinkles skin of chin & protrudes lower lip
Nerve Mandibular br of facial N (VII)

8

MUSCULUS UVULAE
Arises Post border of hard palate
Inserts Palatine aponeurosis
Action Shapes uvula
Nerve Pharyngeal br of vagus N (X) with its motor fibres from cranial accessory N (XI)

MYLOHYOID
Arises Mylohyoid line on int aspect of mandible
Inserts Ant three quarters: midline raphe. Post quarter: sup border of body of hyoid bone
Action Elevates hyoid bone, supports & raises floor of mouth. Aids in mastication & swallowing
Nerve Mylohyoid N (Vc)

NASALIS (COMPRESSOR & DILATOR)
Arises Frontal process of maxilla
Inserts Nasal aponeurosis
Action Opens & closes nostrils, particulary in forced respiration
Nerve Buccal br of facial N (VI')
Notes Part of dilator nasalis is depressor septi from maxilla above central incisor to mobile part of nasal septum

OBLIQUE ARYTENOID
Arises Muscular process of arytenoid cartilage
Inserts Sup pole of opposite arytenoid cartilage
Action Adducts arytenoid cartilages (closes rima glottidis)
Nerve Recurrent laryngeal br of vagus N (X)

OBLIQUUS CAPITIS INFERIOR
Arises Spinous process of axis (C2)
Inserts Lat mass of atlas (C1)
Action Rotates atlanto-axial jnt
Nerve Suboccipital N (post primary ramus of C1)

OBLIQUUS CAPITIS SUPERIOR
Arises Lat mass of atlas (C1)
Inserts Lat half inf nuchal line
Action Lat flexes atlanto-occipital jnt
Nerve Suboccipital N (post primary ramus of C1)

OBTURATOR EXTERNUS
Arises Outer obturator membrane, rim of pubis & ischium bordering it
Inserts Trochanteric fossa on med surface of greater trochanter
Action Lat rotates hip
Nerve Post div of obturator N (L2,3,4)

OBTURATOR INTERNUS
Arises Inner surface of obturator membrane & rim of pubis & ischium bordering membrane
Inserts Middle part of med aspect of greater trochanter of femur
Action Lat rotates & stabilises hip
Nerve N to obturator internus (L5,S1,2)

OCCIPITOFRONTALIS
Arises Occipital: highest nuchal line & mastoid process. Frontal: sup fibres of upper facial muscles
Inserts Galeal aponeurosis
Action Wrinkles forehead & fixes galeal aponeurosis
Nerve Post auricular & temporal brs of facial (VII)

OMOHYOID
Arises Suprascapular lig & adjacent scapula
Inserts Inf border of body of hyoid bone
Action Depresses hyoid bone & hence larynx
Nerve Ansa cervicalis N (C1,2,3)
Notes Tendon between two bellies through sling behind sternocleidomastoid

OPPONENS DIGITI MINIMI (hand)
Arises Flexor retinaculum & hook of hamate

8

Inserts Ulnar border of shaft of 5th MC

Action Opposes (flexes & lat rotates) CMC & MCP jnts of little finger

Nerve Deep br of ulnar N (C8,T1)

OPPONENS POLLICIS

Arises Flexor retinaculum & tubercle of trapezium

Inserts Whole of radial border of 1st MC

Action Opposes (med rotates & flexes) CMC & MCP jnts of thumb

Nerve Recurrent (muscular) br of median N (C8,T1) (may also be from deep br of ulnar N (T1))

ORBICULARIS OCULI

Arises Med orbital margin & lacrimal sac (orbital, palpebral & lacrimal parts)

Inserts Lat palpebral raphe

Action Closes eyelids, aids passage & drainage of tears

Nerve Temporal & zygomatic brs of facial N (VII)

ORBICULARIS ORIS

Arises Near midline on ant surface of maxilla & mandible & modiolus at angle of mouth

Inserts Mucous membrane of margin of lips & raphe with buccinator at modiolus

Action Narrows orifice of mouth, purses lips & puckers lip edges

Nerve Buccal br of facial N (VII)

Notes Accessory parts are incisivus labii superioris & inferioris

PALATOGLOSSUS

Arises Palatine aponeurosis

Inserts Posterolateral tongue

Action Elevates post tongue & closes oropharyngeal isthmus & aids initiation of swallowing

Nerve Pharyngeal br of vagus N (X) with its motor fibres from cranial accessory N (XI)

PALATOPHARYNGEUS

Arises Palatine aponeurosis & post margin of hard palate

Inserts Upper border of thyroid cartilage & blends with constrictor fibres. Upper fibres interdigitate with opposite side (Passavant's ridge)

Action Elevates pharynx & larynx. Passavant's muscle closes nasopharyngeal isthmus in swallowing

Nerve Pharyngeal br of vagus N (X) with its motor fibres from cranial accessory N (XI)

PALMARIS BREVIS

Arises Flexor retinaculum & palmar aponeurosis

Inserts Skin of palm into dermis

Action Steadies & corrugates skin of palm to help with grip

Nerve Superficial br of ulnar N (C8,T1)

Notes Only muscle supplied by this br of ulnar N

PALMARIS LONGUS

Arises Common flexor origin of med epicondyle of humerus

Inserts Flexor retinaculum & palmar aponeurosis

Action Flexes wrist & tenses palmar aponeurosis

Nerve Median N (C7,8) (from med & lat cords)

Notes Absent in 13%

PECTINEUS

Arises Pectineal line of pubis & narrow area of sup pubic ramus below it

Inserts A verticle line between spiral line & gluteal crest below lesser trochanter of femur

Action Flexes, adducts & med rotates hip

Nerve Ant div of femoral N (L2,3). Occasional twig from obturator N (ant div − L2,3)

8

PECTORALIS MAJOR
Arises Clavicular head — med half clavicle. Sternocostal head — lat manubrium & sternum, six upper costal cartilages & ext oblique aponeurosis
Inserts Lat lip of bicipital groove of humerus and ant lip of deltoid tuberosity
Action Clavicular head: flexes & adducts arm. Sternal head: adducts & med rotates arm. Accessory for inspiration
Nerve Med pectoral N (from med cord) & lat pectoral N (from lat cord) (C6,7,8)
Notes Muscle folds on itself so that clavicular fibres insert lowest. Sternal fibres are highest inserting into capsule of shoulder joint

PECTORALIS MINOR
Arises 3, 4, 5 ribs
Inserts Med & upper surface of coracoid process of scapula
Action Elevates ribs if scapula fixed, protracts scapula (assists serratus ant)
Nerve Med & lat pectoral Ns (C6,7,8) (from med & lat cords)
Notes Landmark for axillary art & cords of brachial plexus

PERONEUS BREVIS
Arises Lower two thirds lat shaft of fibula
Inserts Tuberosity of base of 5th MT
Action Dorsiflexes & everts foot. Supports lat longitudinal arch
Nerve Superficial peroneal N (L5,S1)

PERONEUS LONGUS
Arises Upper two thirds of lat shaft of fibula, head of fibula & sup tibiofibular jnt
Inserts Plantar aspect of base of 1st MT & med cuneiform, passing deep to long plantar lig
Action Plantar flexes & everts foot. Supports lat longitudinal & transverse arches
Nerve Superficial peroneal N (L5,S1)

PERONEUS TERTIUS
Arises Third quarter of ant shaft of fibula
Inserts Shaft & base of 5th MT
Action Extends & everts foot
Nerve Deep peroneal N (L5,S1)

PIRIFORMIS
Arises 2, 3, 4 costotransverse bars of ant sacrum, few fibres from sup border of greater sciatic notch
Inserts Ant part of med aspect of greater trochanter of femur
Action Lat rotates & stabilises hip
Nerve Ant primary rami of S1,2
Notes Passes lat through greater sciatic foramen

PLANTARIS
Arises Lat supracondylar ridge of femur above lat head of gastrocnemius
Inserts Tendo calcaneus (med side, deep to gastrocnemius tendon)
Action Plantar flexes foot & flexes knee
Nerve Tibial N (S1,2)

PLATYSMA
Arises Skin over lower neck & upper lat chest
Inserts Inf border of mandible & skin over lower face & angle of mouth
Action Depresses & wrinkles skin of lower face & mouth. Aids forced depression of mandible
Nerve Cervical br of facial N (VII)

POPLITEUS
Arises Post shaft of tibia above soleal line & below tibial condyles
Inserts Middle of three facets on lat surface of lat condyle of femur. Tendon passes into capsule of knee to post part of lat meniscus
Action Unlocks extended knee by lat rotation of femur on tibia. Pulls back lat meniscus
Nerve Tibial N (L5, S1)
Notes Popliteus bursa lies deep to tendon

8

POSTERIOR CRICOARYTENOID
Arises Post aspect of cricoid cartilage
Inserts Muscular process of arytenoid cartilage
Action Abducts & lat rotates arytenoid cartilage (opens rima glottidis)
Nerve Recurrent laryngeal br of vagus N (X)

PROCERUS
Arises Nasal bone & cartilages
Inserts Skin of med forehead
Action Wrinkles & 'frowns' forehead
Nerve Temporal br of facial N (VII)

PRONATOR QUADRATUS
Arises Lower quarter of anteromedial shaft of ulna
Inserts Lower quarter of anterolateral shaft of radius & some interosseous membrane
Action Pronates forearm & maintains ulna & radius opposed
Nerve Ant interosseous N (C8)

PRONATOR TERES
Arises Humeral head: med epicondyle, med supracondylar ridge & med intermuscular septum. Ulnar head: med border of coronoid process
Inserts Just post to most prominent part of lat convexity of radius
Action Pronates forearm & flexes elbow
Nerve Median N (C6,7) (from lat & med cords)
Notes Median N passes between its two heads

PSOAS MAJOR
Arises Transverse processes of L1−5, bodies of T12−L5 & intervertebral discs below bodies of T12−L4
Inserts Middle surface of lesser trochanter of femur
Action Flexes & med rotates hip
Nerve Ant primary rami of L1,2

PSOAS MINOR
Arises Bodies of T12 & L1 & intervening intervertebral disc
Inserts Fascia over psoas major & iliacus
Action Weak flexor of trunk
Nerve Ant primary rami of L1

PYRAMIDALIS
Arises Pubic crest ant to origin of rectus abdominis
Inserts Lower linea alba
Action Reinforces lower rectus sheath
Nerve Subcostal N (T12)

QUADRATUS FEMORIS
Arises Lat border of ischial tuberosity
Inserts Quadrate tubercle of femur & a vertical line below this to the level of lesser trocanter
Action Lat rotates & stabilises hip
Nerve N to quadratus femoris (L4,5,S1)

QUADRATUS LUMBORUM
Arises Inf border of 12th rib
Inserts Apices of transverse processes of L1−4, iliolumbar lig & post third of iliac crest
Action Fixes 12th rib during respiration & lat flexes trunk
Nerve Ant primary rami (T12−L3)

RECTUS ABDOMINIS
Arises Pubic crest & pubic symphysis
Inserts 5, 6, 7 costal cartilages, med inf costal margin & post aspect of xiphoid
Action Flexes trunk, aids forced expiration & raises intra-abdominal pressure
Nerve Ant primary rami (T7−12)

RECTUS CAPITIS ANTERIOR
Arises Lat mass of atlas (C1)
Inserts Basilar occipital bone ant to occipital condyle
Action Flexes atlanto-occipital jnt
Nerve Ant primary rami of C1

8

RECTUS CAPITIS LATERALIS
Arises Lat mass of atlas (C1)
Inserts Jugular process of occipital bone
Action Lat flexes atlanto-occipital jnt
Nerve Ant primary rami of C1

RECTUS CAPITIS POSTERIOR MAJOR
Arises Spinous process of axis (C2)
Inserts Lat half of inf nuchal line
Action Extends & rotates atlanto-occipital jnt
Nerve Suboccipital N (post primary ramus C1)

RECTUS CAPITIS POSTERIOR MINOR
Arises Post process of atlas (C1)
Inserts Med half of inf nuchal line
Action Extends atlanto-occipital jnt
Nerve Suboccipital N (post primary ramus C1)

RECTUS FEMORIS (QUADRICEPS FEMORIS I)
Arises Straight head: ant inf iliac spine. Reflected head: ilium above acetabulum
Inserts Quadriceps tendon to patella, via ligamentum patellae into tubercle of tibia
Action Extends leg at knee. Flexes thigh at hip
Nerve Post div of femoral N (L3,4)

RHOMBOID MAJOR
Arises Spines of T2–T5 & supraspinous ligs
Inserts Lower half of posteromedial border of scapula, from angle to upper part of triangular area at base of scapular spine
Action Retracts scapula. Rotates scapula to rest position
Nerve Dorsal scapular N (C5) (from root)

RHOMBOID MINOR
Arises Lower ligamentum nuchae, spines of C7 & T1

Inserts Small area of posteromedial border of scapula at level of spine, below levator scapulae
Action Retracts scapula. Rotates lower scapula back to rest position
Nerve Dorsal scapular N (C5) (from root)

RISORIUS
Arises Deep fascia of face & parotid
Inserts Modiolus & skin at angle of mouth
Action Retracts angle of mouth
Nerve Buccal br of facial N (VII)

SALPINGOPHARYNGEUS
Arises Inf cartilage & mucosa of pharyngeal orifice of auditory tube
Inserts Upper border of thyroid cartilage & inf constrictor muscle fibres
Action Elevates pharynx & larynx & aids swallowing. Opens auditory canal during swallowing
Nerve Pharyngeal br of vagus N (X) with its motor fibres from cranial accessory N (XI)

SARTORIUS
Arises Immediately below ant sup iliac spine
Inserts Upper med surface of shaft of tibia
Action Flexes, abducts, lat rotates thigh at hip. Flexes, med rotates leg at knee
Nerve Ant div of femoral N (L3,4)

SCALENUS ANTERIOR
Arises Ant tubercles of transverse processes of C3–6
Inserts Scalene tubercle on sup aspect of 1st rib
Action Accessory to inspiration. Lat flexion of neck when 1st rib fixed
Nerve Ant primary rami of C5,6

SCALENUS MEDIUS
Arises Post tubercles of transverse processes of C2–7
Inserts Sup aspect of neck of 1st rib

8

Action Accessory to inspiration
Nerve Ant primary rami of C3−8

SCALENUS MINIMUS
Arises Ant tubercle of transverse process of C7
Inserts Suprapleural membrane (Sibson's fascia)
Action Supports suprapleural membrane
Nerve Ant primary rami of C7

SCALENUS POSTERIOR
Arises Post tubercles of transverse processes C4−6
Inserts Post/lat surface of 2nd rib
Action Accessory to inspiration
Nerve Ant primary rami of C6−8

SEMIMEMBRANOSUS
Arises Upper outer quadrant of post surface of ischial tuberosity
Inserts Med condyle of tibia below articular margin, fascia over popliteus & oblique popliteal lig
Action Flexes & med rotates knee. Extends hip
Nerve Tibial portion of sciatic N (L5,S1)

SEMITENDINOSUS
Arises Upper inner quadrant of post surface of ischial tuberosity
Inserts Upper med shaft of tibia below gracilis
Action Flexes & med rotates knee. Extends hip
Nerve Tibial portion of sciatic N (L5,S1)

SERRATUS ANTERIOR
Arises Upper eight ribs & ant intercostal membranes from midclavicular line. Lower four interdigitating with external oblique
Inserts Inner med border scapula. 1 & 2: upper angle; 3 & 4: length of costal surface; 5−8: inf angle
Action Lat rotates & protracts scapula

Nerve Long thoracic N of Bell (C5,6,7) (from roots) slips from ribs 1 & 2: C5; 3 & 4: C6; 5−8: C7

SERRATUS POSTERIOR INFERIOR
Arises Spinous processes & supraspinous ligs of T11−L2
Inserts Post aspect of ribs 9−12
Action Assists forced expiration
Nerve Ant primary rami (T9−12)

SERRATUS POSTERIOR SUPERIOR
Arises Spinous processes & supraspinous lig of C7−T2
Inserts Post aspect of ribs 2−5
Action Assists forced inspiration
Nerve Ant primary rami (T2−5)

SOLEUS
Arises Soleal line & middle third of post border of tibia & upper quarter of post shaft of fibula including neck
Inserts Tendo calcaneus to middle of three facets on post surface of calcaneus
Action Plantar flexes foot (aids venous return)
Nerve Tibial N (S1,2)
Notes Main propulsive force for walking & running

SPHINCTER ANI (external)
Arises Circular anatomical sphincter
Inserts Deep, superficial & subcutaneous portions
Action Maintains continence of faeces
Nerve Inf rectal br of pudendal N (S2,3,4)

SPHINCTER URETHRAE
Arises Circular anatomical sphincter
Inserts Fuses with deep transverse perinei
Action Maintains continence of urine
Nerve Perineal br of pudendal N (S2,3,4)

SPLENIUS CAPITIS
Arises Lower lig nuchae, spinous processes & supraspinous ligs T1−3
Inserts Lat occiput between sup & inf nuchal lines

8

Action Extends & rotates cervical spine
Nerve Post primary rami of C3,4

SPLENIUS CERVICIS

Arises Spinous processes & supraspinous ligs of T3−6
Inserts Post tubercles of transverse processes of C1−3
Action Extends & rotates cervical spine
Nerve Post primary rami of C5,6

STAPEDIUS

Arises The pyramid (post wall of middle ear)
Inserts Neck of stapes
Action Protects & critically damps ossicular chain
Nerve Facial (VII), in middle ear

STERNOCLEIDOMASTOID

Arises Ant & sup manubrium & sup med third of clavicle
Inserts Lat aspect of mastoid process & ant half of sup nuchal line
Action Flexes & lat rotates cervical spine. Protracts head when acting together. Extends neck when neck already partially extended
Nerve Spinal accessory N (lat roots C1−5)
Notes Effectively four separate muscles

STERNOHYOID

Arises Sup lat post aspect of manubrium
Inserts Inf border of body of hyoid bone
Action Depresses hyoid bone & hence larynx
Nerve Ansa cervicalis N (C1,2,3)

STERNOTHYROID

Arises Med post aspect of manubrium
Inserts Oblique line of lamina of thyroid cartilage
Action Depresses larynx
Nerve Ansa cervicalis N (C1,2,3)

STYLOGLOSSUS

Arises Ant surface & apex of styloid process & upper quarter of stylohyoid lig
Inserts Superolateral sides of tongue
Action Retracts & elevates tongue, aids initiation of swallowing
Nerve Hypoglossal N (XII)

STYLOHYOID

Arises Base of styloid process
Inserts Base of greater cornu of hyoid bone
Action Elevates & retracts hyoid bone. Aids swallowing & elevates larynx
Nerve Mandibular br of facial N (VII)

STYLOPHARYNGEUS

Arises Med aspect of styloid process
Inserts Posterolateral border of thyroid cartilage
Action Elevates larynx & pharynx. Aids swallowing
Nerve Muscular br of glossopharyngeal N (IX)

SUBCLAVIUS

Arises Costochondral junction of 1st rib
Inserts Subclavian groove on inf surface of middle third of clavicle
Action Depresses clavicle & steadies it during shoulder movements
Nerve N to subclavius (C5,6, upper trunk)

SUBCOSTALIS

Arises Int post aspects of lower six ribs
Inserts Int aspects of ribs two to three levels below
Action Depresses lower ribs
Nerve Muscular collateral brs of intercostal Ns

SUBSCAPULARIS

Arises Med two thirds of subscapular fossa
Inserts Lesser tuberosity of humerus, upper med lip of bicipital groove, capsule of shoulder jnt
Action Med rotates arm & stabilises shoulder jnt

8

Nerve Upper & lower subscapular Ns (C6,7) (from post cord)

Notes Subscapular bursa beneath tendon, usually connected with jnt. Tendon forms part of rotator cuff

SUPERFICIAL TRANSVERSE PERINEI

Arises Body of ischium

Inserts Perineal body

Action Fixes perineal body

Nerve Perineal br of pudendal N (S2,3,4)

SUPERIOR OBLIQUE (see also obliquus capitis superior)

Arises Body of sphenoid above tendinous ring

Inserts Post/sup quadrant of sclera behind equator of eyeball

Action Depresses eye in adduction. Intorts eye in abduction

Nerve Trochlear N (IV)

Notes Passes around trochlear sling on frontal bone

SUPERIOR RECTUS

Arises Sup tendinous ring within orbit

Inserts Superior sclera ant to equator of eyeball

Action Elevates eye. Intorts eye in adduction

Nerve Sup div of oculomotor N (III)

SUPINATOR

Arises Deep part (horizontal): supinator crest & fossa of ulna. Superficial part (downwards): lat epicondyle & lat lig of elbow & annular lig

Inserts Neck & shaft of radius, between ant & post oblique lines

Action Supinates forearm. Only acts alone when elbow extended

Nerve Post interosseous N (C5,6)

Notes Post interosseous N passes between two heads

SUPRASPINATUS

Arises Med three quarters of supraspinous fossa of scapula, upper surface of spine (bipennate)

Inserts Sup facet on greater tuberosity of humerus & capsule of shoulder jnt

Action Abducts arm & stabilizes shoulder jnt

Nerve Suprascapular N (C5,6) (from upper trunk)

Notes Subacromial bursa lies above its tendon. Tendon forms part of rotator cuff

TEMPORALIS

Arises Temporal fossa between inf temporal line & infratemporal crest

Inserts Med & ant aspect of coronoid process of mandible

Action Elevates mandible & post fibres retract

Nerve Deep temporal brs from ant div of mandibular N (Vc)

TEMPOROPARIETALIS

Arises Aponeurosis above auricularis

Inserts Galeal aponeurosis

Action Fixes galeal aponeurosis

Nerve Temporal br of facial N (VII)

TENSOR FASCIAE LATAE

Arises Outer surface of ant iliac crest between tubercle of the iliac crest & ant sup iliac spine

Inserts Iliotibial tract (ant surface of lat condyle of tibia)

Action Maintains knee extended (assists gluteus maximus) & abducts hip

Nerve Sup gluteal N (L4,5,S1)

TENSOR TYMPANI

Arises Cartilaginous & bony margins of auditory tube

Inserts Handle of malleus (via processus trochleariformis)

Action Protects & critically damps ossicular chain

Nerve N to med pterygoid (main trunk of mandibular N (Vc))

8

TENSOR VELI PALATINI
Arises Scaphoid fossa & med aspect of spine of sphenoid bone
Inserts Palatine aponeurosis (via pulley of pterygoid hamulus) .
Action Tenses soft palate prior to elevation
Nerve N to med pterygoid (main trunk of mandibular N (Vc))

TERES MAJOR
Arises Oval area (lower third) of lat side of inf angle of scapula below teres minor
Inserts Med lip of bicipital groove of humerus
Action Med rotates & adducts arm. Stabilises shoulder jnt
Nerve Lower subscapular N (C5,6) (from post cord)
Notes Functionally part of subscapularis

TERES MINOR
Arises Middle third lat border of scapula above teres major
Inserts Inf facet of greater tuberosity of humerus (below infraspinatus) & capsule of shoulder jnt
Action Lat rotates arm & stabilises shoulder jnt
Nerve Axillary N (C5,6) (from post cord)
Notes Functionally part of infraspinatus. Tendon forms part of rotator cuff

THYRO-ARYTENOID & VOCALIS
Arises Lower post surface of angle between laminae of thyroid cartilage
Inserts Vocal process of arytenoid cartilage
Action Shortens & relaxes vocal cords by med rotation & protraction of arytenoid cartilage
Nerve Recurrent laryngeal br of vagus N (X)

THYRO-EPIGLOTTICUS
Arises Lower post surface of thyroid cartilage
Inserts Lat border of epiglottis

Action Aids closure of additus to larynx
Nerve Recurrent laryngeal br of vagus N (X)

THYROHYOID
Arises Oblique line of lamina of thyroid cartilage
Inserts Inf border of body of hyoid bone
Action Elevates larynx or depresses hyoid bone
Nerve C1 fibres carried by hypoglossal N (XII)

TIBIALIS ANTERIOR
Arises Upper half of lat shaft of tibia & interosseous membrane
Inserts Inferomedial aspect of med cuneiform & base of 1st MT
Action Extends & inverts foot at ankle. Holds up med longitudinal arch of foot
Nerve Deep peroneal N (L4,5)
Notes Inversion is at subtalar & transverse tarsal joints

TIBIALIS POSTERIOR
Arises Upper half of post shaft of tibia & upper half of fibula between median crest & interosseous border, & interosseous membrane
Inserts Tuberosity of navicular bone & all tarsal bones (except talus) & spring lig
Action Plantar flexes & inverts foot. Supports med longitudinal arch of foot
Nerve Tibial N (L4,5)

TRANSVERSE ARYTENOID
Arises Post surface & muscular process of arytenoid cartilage
Inserts Corresponding surfaces of opposite cartilage
Action Adducts arytenoid cartilage (closes rima glottidis)
Nerve Recurrent laryngeal branch of vagus N (X)

8

TRANSVERSOSPINALIS — MULTIFIDUS
Arises Laminae of vertebra
Inserts Spinous processes two to three levels above
Action Extends spine
Nerve Post primary rami

TRANSVERSOSPINALIS — ROTATORES
Arises Transverse processes
Inserts Spinous processes one above
Action Rotates spine
Nerve Post primary rami
Notes Two types — thoracis & cervicis et lumborum

TRANSVERSOSPINALIS — SEMISPINALIS
Arises Transverse processes
Inserts Spinous processes six levels above
Action Extends & lat flexes spine
Nerve Post primary rami
Notes Three types — thoracis, cervicis & capitis

TRANSVERSUS ABDOMINIS
Arises Costal margin, lumbar fascia, ant two thirds of iliac crest & lat half of inguinal lig
Inserts Aponeurosis of post & ant rectus sheath & conjoint tendon to pubic crest & pectineal line
Action Supports abdominal wall, aids forced expiration & raising intra-abdominal pressure. Conjoint tendon supports post wall of inguinal canal
Nerve Ant primary rami (T7–12). Conjoint tendon ilioinguinal N (L1)

TRANSVERSUS THORACIS (STERNOCOSTALIS)
Arises Lower third of inner aspect of sternum & lower three costosternal junctions
Inserts Second to sixth costal cartilages
Action Depresses upper ribs

Nerve Muscular collateral brs of intercostal Ns

TRAPEZIUS
Arises Med third sup nuchal line, lig nuchae, spinous processes & supraspinous ligs to T12
Inserts Upper fibres to lat third of post border of clavicle; lower to med acromion & sup lip of spine of scapula to deltoid tubercle
Action Lat rotates, elevates & retracts scapula. If scapula is fixed, extends & lat flexes neck
Nerve Spinal accessory N (lat roots, C1–5) (spinal nerves C3 & C4 for proprioception)

TRICEPS
Arises Long head: infraglenoid tubercle of scapula. Lat head: upper half post humerus (linear origin). Med head: lies deep on lower half post humerus inferomedial to spiral groove & both intermuscular septa
Inserts Post part of upper surface of olecranon process of ulna & post capsule
Action Extends elbow. Long head stabilises shoulder jnt. Med head retracts capsule of elbow jnt on extension
Nerve Radial N (C7,8) (from post cord), four brs

VASTUS INTERMEDIUS (QUADRICEPS FEMORIS 2)
Arises Ant & lat shaft of femur to one hand's breadth above condyles
Inserts Quadriceps tendon to patella, via ligamentum patellae into tubercle of tibia
Action Extends knee
Nerve Post div of femoral N (L3,4)

VASTUS LATERALIS (QUADRICEPS FEMORIS 3)
Arises Upper intertrochanteric line, base of greater trochanter, lat linea aspera, lat supracondylar ridge & lat intermuscular septum

8

Inserts Lat quadriceps tendon to patella, via ligamentum patellae into tubercle of tibia
Action Extends knee
Nerve Post div of femoral N (L3,4)

VASTUS MEDIALIS (QUADRICEPS FEMORIS 4)
Arises Lower intertrochanteric line, spiral line, med linea aspera & med intermuscular septum
Inserts Med quadriceps tendon to patella & directly into med patella, via ligamentum patellae into tubercle of tibia
Action Extends knee. Stabilises patella
Nerve Post div of femoral N (L3,4)

ZYGOMATICUS MAJOR
Arises Ant surface of zygomatic bone
Inserts Modiolus at angle of mouth
Action Elevates & draws angle of mouth laterally
Nerve Buccal br of facial N (VII)

ZYGOMATICUS MINOR
Arises Lat infra-orbital margin
Inserts Skin & muscle of upper lip
Action Elevates & everts upper lip
Nerve Buccal br of facial N (VII)

8

9: JOINTS

Classification of joints

Fibrous Fibrous tissue between bone ends
Primary cartilaginous Hyaline cartilage between bone ends
Secondary cartilaginous As for primary but fibrocartilage between the layers of hyaline cartilage (symphysis)
Synovial Joint cavity with synovial fluid. Hyaline cartilage on surface of bones. Articular disc can be present
Atypical synovial Joint cavity with synovial fluid. Fibrocartilage on surface of bones. Articular disc can be present

Types of synovial joint

Plane No degrees of freedom. Sliding only
Hinge (ginglymi) One plane of movement
Modified hinge (bicondylar) One plane of movement + rotation
Condyloid (ellipsoid) Two planes of movement (circumduction)
Saddle condyloid (sella) Two planes of movement + rotation
Pivot (trochoid) Rotation only
Ball and socket (spheroidal) Multi-axial
Gomphosis Peg and socket joint. No movement

Joints with interarticular fibrocartilaginous discs

Acromioclavicular (usually incomplete)
Femorotibial (knee) (incomplete — menisci)
Radiocarpal (wrist)
Sternoclavicular
Temporomandibular

Joints with double cavities separated by intra-articular ligaments — not fibrocartilaginous discs

Costovertebral (ribs 2−10)
Sternochondral (2nd rib)

Joints classified by type

Fibrous joints
Arytenocorniculate (can be synovial)
Costotransverse (ribs 11 and 12)
Cuboideonavicular (can be synovial)
Gomphosis (teeth)
Radio-ulnar (interosseous membrane)
Skull sutures
(stylohyoid ligament)
Tibiofibular (inferior)
Tibiofibular (interosseous membrane)

Primary cartilaginous joints
Costochondral
Sternochondral (1st rib)

Secondary cartilaginous joints
Intervertebral
Manubriosternal
Sacrococcygeal
Symphysis pubis
Xiphisternal

Atypical synovial joints
Acromioclavicular
Sternochondral (ribs 2−7)
Sternoclavicular
Temporomandibular
Sacro-iliac (partially atypical)

Typical synovial
Acetabulofemoral (hip)
Atlanto-axial (dens & facets)
Atlanto-occipital
Calcaneocuboid
Carpometacarpal
Costotransverse (ribs 1−10)
Costovertebral
Crico-arytenoid
Crico-thyroid
Cuneocuboid
Cuneonavicular
Femorotibial (knee)
Glenohumeral (shoulder)
Humeroradial (elbow)

9

Humero-ulnar (elbow)
Intercarpal
Interchondral (cartilages 6−7,7−8,8−9)
Intercuneiform
Intermetacarpal
Intermetatarsal
Interphalangeal
Metacarpophalangeal
Metatarsophalangeal
Pisotriquetral
Radiocarpal (wrist)
Radio-ulnar (superior & inferior)
Talocalcaneal
Talocalcaneonavicular
Tarsometatarsal
Tibiofibular (superior)
Tibiotalal (ankle)
Zygapophyseal (intervertebral facet)

Unclassified
Intervertebral joints of Luschka

9

9

ACETABULOFEMORAL (hip)
Classification Synovial
Type Ball & socket
Articulation Acetabulum with femur

ACROMIOCLAVICULAR
Classification Atypical synovial
Type Plane
Articulation Acromion with clavicle
Notes Often an articular disc in upper part of jnt, usually incomplete

ANKLE (see tibiotalal)

ARYTENOCORNICULATE (larynx)
Classification Fibrous or synovial
Articulation Arytenoid cartilage with corniculate cartilage

ATLANTO-AXIAL — LATERAL
Classification Synovial
Type Plane
Articulation Articular facets of atlas with axis

ATLANTO-AXIAL — MEDIAN
Classification Synovial
Type Pivot
Articulation Dens of axis with atlas
Notes Second cavity (bursa) posteriorly

ATLANTO-OCCIPITAL
Classification Synovial
Type Condyloid
Articulation Atlas with occipital bone

CALCANEOCUBOID (midtarsal)
Classification Synovial
Type Saddle condyloid
Articulation Calcaneus with cuboid

CARPOMETACARPAL — FINGERS
2 – 5 (including intermetacarpal)
Classification Synovial

Type Plane
Articulation Carpal bones with MCs & between MCs
Notes Usually continuous cavity between CMC, intermetacarpal & intercarpal jnts

CARPOMETACARPAL — THUMB
Classification Synovial
Type Saddle condyloid
Articulation Trapezium with 1st MC
Notes Joint is separate from others in hand

COSTOCHONDRAL
Classification Primary cartilaginous
Articulation Bony rib with costal cartilage

COSTOTRANSVERSE — RIBS 1 – 10
Classification Synovial
Type Plane
Articulation Med facet of tubercle of rib with transverse process of own vertebra

COSTOTRANSVERSE — RIBS 11, 12
Classification Ligamentous (fibrous)
Articulation Tubercle of rib with transverse process of own vertebra

COSTOVERTEBRAL
Classification Synovial
Type Plane
Articulation Head of rib with vertebral body
Notes 1st rib articulates with T1 vertebra only (single cavity jnt). Ribs 2 – 10 with own vertebra & one above (double cavity jnts separated by intra-articular lig). Ribs 11 & 12 with own vertebra only (single cavity jnts)

CRICO-ARYTENOID (larynx)
Classification Synovial
Type Ball & socket
Articulation Cricoid cartilage with arytenoid cartilage

9

CRICOTHYROID (larynx)
Classification Synovial
Type Hinge
Articulation Facet on side of cricoid
   cartilage with inf horn of thyroid cartilage

ELBOW (see humeroradial & humero-ulnar)

FEMOROTIBIAL (knee)
Classification Synovial
Type Modified hinge
Articulation Femur with tibia
Notes Menisci are incomplete discs of
   fibrocartilage

GLENOHUMERAL (shoulder)
Classification Synovial
Type Ball & socket
Articulation Glenoid fossa of scapula with
   humerus

GOMPHOSIS (dento-alveolar)
Classification Fibrous
Articulation Tooth with bone of jaw

HIP (see acetabulofemoral)

HUMERORADIAL (elbow)
Classification Synovial
Type Hinge
Articulation Capitulum of humerus with
   radial head
Notes Joint cavity is shared with humero-
   ulnar & sup radio-ulnar jnts

HUMERO-ULNAR (elbow)
Classification Synovial
Type Hinge
Articulation Trochlea of humerus with
   ulnar trochlear notch
Notes Joint cavity is shared with
   humeroradial & sup radio-ulnar jnts

INTERCARPAL (MIDCARPAL) (see also
pisotriquetral)
Classification Synovial
Type Plane individually but together give
   effective mixture of condyloid, saddle
   condyloid & ball & socket
Articulation Between scaphoid, lunate,
   triquetral, hamate, capitate, trapezoid &
   trapezium
Notes Single cavity between the seven
   bones usually communicating also with
   CMC & intercarpal jnts of fingers 2−5

INTERCHONDRAL
Classification Synovial
Type Plane
Articulation Between costal cartilages 6/7,
   7/8, 8/9

INTERMETATARSAL
Classification Synovial
Type Plane
Articulation Between MTs

INTERPHALANGEAL (fingers & toes)
Classification Synovial
Type Hinge
Articulation Between phalanges

**INTERTARSAL − CUBOIDEO-
NAVICULAR**
Classification Fibrous (can be synovial)
Articulation Cuboid with navicular

INTERTARSAL − CUNEOCUBOID
Classification Synovial
Type Plane
Articulation Lat cuneiform with cuboid
Notes Cuneocuboid shares cavity with
   cuneonavicular & intercuneiform jnts

INTERTARSAL-CUNEONAVICULAR
Classification Synovial
Type Plane
Articulation Cuneiforms with navicular

9

Notes Cuneonavicular shares cavity with cuneocuboid and intercuneiform jnts

INTERTARSAL — INTERCUNEIFORM
Classification Synovial
Type Plane
Articulation Between cuneiforms
Notes Intercuneiform jnts share cavity with cuneonavicular & cuneocuboid jnts

INTERVERTEBRAL
Classification Secondary cartilaginous
Articulation Between vertebral bodies

INTERVERTEBRAL JOINTS OF LUSCHKA (neurocentral or uncovertebral)
Classification Unclassified
Type Unclassified
Articulation Posterolateral lip of upper surface of C3−7 & T1 vertebrae with adjacent disc
Notes Often small cavity which is degenerative (not synovial)

KNEE (see femorotibial)

9

MANUBRIOSTERNAL
Classification Secondary cartilaginous
Articulation Manubrium with sternum
Notes May cavitate to give appearance of synovial jnt

METACARPOPHALANGEAL
Classification Synovial
Type Condyloid
Articulation MCs with phalanges

METATARSOPHALANGEAL
Classification Synovial
Type Condyloid
Articulation MTs with phalanges

PISOTRIQUETRAL
Classification Synovial
Type Plane
Articulation Pisiform with triquetral

RADIOCARPAL (wrist)
Classification Synovial
Type Condyloid
Articulation Radius & triangular fibro-cartilaginous articular disc with scaphoid, lunate & triquetral

RADIO-ULNAR — INFERIOR
Classification Synovial
Type Pivot
Articulation Radius with ulna
Notes Cavity separated from cavity of wrist by triangular fibrocartilaginous disc

RADIO-ULNAR — INTEROSSEOUS MEMBRANE & OBLIQUE CORD
Classification Fibrous
Articulation Radius with ulna

RADIO-ULNAR — SUPERIOR
Classification Synovial
Type Pivot
Articulation Radius with ulna
Notes Cavity is continuous with humero-ulnar & humeroradial jnts

SACROCOCCYGEAL
Classification Secondary cartilaginous
Articulation Sacrum with coccyx

SACRO-ILIAC
Classification Partially atypical synovial
Type Plane
Articulation Sacrum with iliac bone
Notes Has fibrocartilage on ilium

SHOULDER (see glenohumeral)

SKULL SUTURES
Classification Fibrous
Articulation Between diploae of skull

STERNOCHONDRAL (sternocostal)
Classification 1st: primary cartilaginous;
2nd−7th: atypical synovial
Type Plane
Articulation 1st rib with manubrium.
3rd−7th ribs with sternum. 2nd with both
Notes 2nd jnt has two cavities separated by
intra-articular lig

STERNOCLAVICULAR
(manubrioclavicular)
Classification Atypical synovial
Type Ball & socket
Articulation Clavicle with manubrium
Notes Separated into two cavities by
fibrocartilaginous disc

SYMPHYSIS PUBIS
Classification Secondary cartilaginous
Articulation Between pubic bones
Notes May cavitate

TALOCALCANEAL (subtalar)
Classification Synovial
Type Plane (effectively ball & socket)
Articulation Talus with calcaneus

TALOCALCANEONAVICULAR
Classification Synovial
Type Ball & socket
Articulation Talus with calcaneus &
navicular
Notes Talocalcaneal part is part of subtalar
jnt. Talonavicular part is part of midtarsal
jnt.

TARSOMETATARSAL
Classification Synovial

Type Plane
Articulation Tarsal bones with MTs

TEMPOROMANDIBULAR
Classification Atypical synovial
Type Condyloid
Articulation Temporal bone with mandible
Notes Separated into two cavities by
fibrocartilaginous disc

TIBIOFIBULAR — INFERIOR
Classification Fibrous
Articulation Tibia with fibula

TIBIOFIBULAR — INTEROSSEOUS
MEMBRANE
Classification Fibrous
Articulation Tibia with fibula

TIBIOFIBULAR — SUPERIOR
Classification Synovial
Type Plane
Articulation Tibia with fibula

TIBIOTALAL (ankle)
Classification Synovial
Type Hinge
Articulation Tibia with talus

WRIST (see radiocarpal)

XIPHISTERNAL
Classification Secondary cartilaginous
Articulation Xiphoid with sternum

ZYGAPOPHYSEAL (intervertebral facet)
Classification Synovial
Type Plane
Articulation Between intervertebral facets

9

9

10: OSSIFICATION TIMES

10

Ossification times

Bones (number if unpaired)	Forms in membrane (M) or cartilage (C)	Centres primary (P) or secondary (S)	Site	Centre appears at:		Fused by
				Gestation (weeks/months)	After birth	
Mandible (1)	M	P	Near mental foramen (each side)	6W		Symphysis menti 1–3Y
Hyoid (1)	C	P	Greater cornu (each side)	8–9M		
	C	S	Body (2 centres)	9M		
	C	S	Lesser cornu (each side)		Puberty	
Occiput (1)	M	P	Squamous (each side)	8W		
	C	P	Lateral (each side)	8W		
	C	P	Basilar	8W		
Sphenoid (1)	M+C	P	Approximately 14 centres	8W–4M		
Temporal	M	P	Squamous	8W		
	M	P	Tympanic	3M		
	C	P	Petromastoid (several centres)	5M		
Parietal	M	P	Near tuberosity (2 centres)	7W		
Frontal (2→1)	M	P	Near each tuberosity (2 centres, one each side)	8W		Metopic suture 2Y
Ethmoid (1)	C	P	Labyrinth (one each side)	5M		
	C	P	Perpendicular plate/crista galli		1Y	
Inf concha	M	P		5M		
Lacrimal	M	P		4M		
Nasal	M	P		3M		
Vomer (1)	M	P	(2 centres)	8W		
Maxilla	M	P	(3 centres)	6–8W		
Palatine	M	P	Perpendicular plate	8W		

10

Bone		Structure		Appears	Fusion
Zygomatic	M		P	8 W	
Ear ossicles	C	Stapes	P	4 M	
	C	Malleus	P	4 M	
	C	Incus	P	4 M	
Scapula	C	Body	P	8 W	
	C	Coracoid process	S	1 Y	15 Y
	C	Subcoracoid	S	Puberty	20 Y
	C	Medial border	S	Puberty	20 Y
	C	Glenoid (lower rim)	S	Puberty	20 Y
	C	Acromion (2 centres)	S	Puberty	20 Y
	C	Inferior angle	S	Puberty	20 Y
Clavicle	M	Medial & lateral (2 centres)	P	5 W	
	M	Sternal end	S	Late teens	20 Y
Humerus (upper end is growing end)	C	Shaft	P	8 W	
	C	Head	S	6 M	Upper epiphysis 18—20 Y
	C	Greater tuberosity	S	2 Y	
	C	Lesser tuberosity	S	5 Y	
	C	Capitulum & lat ridge of trochlea	S	1 Y	
	C	Medial trochlea	S	10 Y	Lower epiphysis 14—16 Y
	C	Medial epicondyle	S	5 Y	
	C	Lateral epicondyle	S	12 Y	
Radius (lower end is growing end)	C	Shaft	P	8 W	
	C	Head	S	4 Y	14—17 Y
	C	Distal end	S	1 Y	17—19 Y
Ulna (lower end is growing end)	C	Shaft	P	8 W	
	C	Olecranon (2 centres)	S	9 Y	14—16 Y
	C	Distal end	S	5 Y	17—18 Y
Carpus	C	Capitate	P	2 M	
	C	Hamate	P	3 M	
	C	Triquetral	P	3 Y	
	C	Lunate	P	4 Y	

10

continued on next page

Ossification times continued

Bones (number if unpaired)	Forms in membrane (M) or cartilage (C)	Centres primary (P) or secondary (S)	Site	Centre appears at:		Fused by
				Gestation (weeks/months)	After birth	
	C	P	Scaphoid		4–5Y	
	C	P	Trapezium		4–5Y	
	C	P	Trapezoid		4–5Y	
	C	P	Pisiform		9–12Y	
Metacarpal (1st)	C	P	Shaft	9W		
	C	S	Base		3Y	15–17Y
Metacarpals (2nd–5th)	C	P	Shaft	9W		
	C	S	Head		2Y	15–19Y
Phalanges (hand)	C	P	Shaft	8–12W		
	C	S	Base		2–4Y	15–18Y
Innominate	C	P	Pubis (superior ramus)	4M		7–8Y
	C	P	Ischium (body)	4M		7–8Y
	C	P	Ilium (above greater sciatic notch)	2M		7–8Y
	C	S	Iliac crest (2 centres)		Puberty	15–25Y
	C	S	Acetabulum (2 centres)		Puberty	15–25Y
	C	S	Anterior superior iliac spine		Puberty	15–25Y
	C	S	Ischial tuberosity		Puberty	15–25Y
	C	S	Pubic symphysis		Puberty	15–25Y
Femur (lower end is growing end)	C	P	Shaft	7W		
	C	S	Greater trochanter		4Y	
	C	S	Lesser trochanter		12Y	
	C	S	Head	9M	6M	14–17Y
	C	S	Distal end			16–18Y
Patella	C	P	(Several centres)		3–6Y	Puberty
	C	S	Superolaterally		6Y	Puberty

10

Tibia (upper end is growing end)	C	P	Shaft	7 W	
	C	S	Plateau	9 M	16–18 Y
	C	S	Distal end	1 Y	15–17 Y
	C	S	Tuberosity	12 Y	13–14 Y
Fibula (upper end is growing end)	C	P	Shaft	8 W	
	C	S	Distal end	1 Y	15–17 Y
	C	S	Head	3–4 Y	17–19 Y
Talus	C	P		6 M	
Calcaneus	C	P		3 M	
	C	S		6–8 Y	14–16 Y
Navicular	C	P		3 Y	
Cuneiform lateral	C	P		1 Y	
Cuneiform medial	C	P	(May have 2 centres)	2 Y	
Cuneiform intermediate	C	P		3 Y	
Cuboid	C	P		9 M	
Metatarsal (1st)	C	P	Shaft	9 W	
	C	S	Base	3 Y	17–20 Y
Metatarsals (2nd–5th)	C	P	Shaft	9 W	
	C	S	Head	3–4 Y	17–20 Y
Phalanges (foot)	C	P	Shaft	9–15 W	
	C	S	Base	2–8 Y	18 Y

continued on next page

Closure of skull sutures

Ant fontinelle: closes 18 M; post fontinelle: closes 6 M – 1 Y

Note: (1) All bones are paired unless otherwise stated. (2) Single centre of ossification unless specified otherwise. (3) Variability of ossification usually a sex difference, females appearing and uniting earlier. (4) Fusion times for epiphyses are given if clinically relevant

10

Ossification times continued — eruption of teeth

Eruption of teeth

	Incisor		Canine	Premolar	Molar
	(upper)	(lower)			
First dentition (months)	7,8	6,9	18		12, 24
Second dentition* (years)	7, 8	7, 8	11	9, 10	6, 12, 18

* Lower teeth erupt slightly earlier.

10

Note: (1) Most smaller emissary veins and meningeal arterial supplies have been omitted. (2) All structures are paired unless otherwise indicated.

11

AQUEDUCT OF THE VESTIBULE

Site In post aspect of petrous temporal bone in post cranial fossa, 1 cm post to int acoustic meatus

Contains Endolymphatic duct & sac, small art & V

CAROTID CANAL

Site In inf surface of petrous temporal bone in middle cranial fossa

Contains Int carotid art enters with sympathetic plexus on it. Int carotid venous plexus connecting cavernous sinus & int jugular vein

CONDYLAR CANAL

Site In lower sigmoid groove in occipital bone in post cranial fossa. Exits at condylar fossa behind condyle (not always present)

Contains Emissary V connecting sigmoid sinus & occipital Vs. Meningeal br of occipital art

CRIBRIFORM FORAMINA

Site In cribriform plate of ethmoid bone in ant cranial fossa

Contains Olfactory filaments & ant ethmoidal N & vessels

FACIAL CANAL

Site In petrous temporal bone leading from int acoustic meatus to stylomastoid foramen

Contains Facial N (VII)

FORAMEN CAECUM (unpaired)

Site Between frontal crest of frontal bone & crista galli of ethmoid bone in ant cranial fossa

Contains Emissary Vs connecting nose & sup sagittal sinus

FORAMEN LACERUM

Site Between sphenoid, apex of petrous temporal & basilar occipital bones in middle cranial fossa

Contains Int carotid art enters behind & exits above. Greater petrosal N enters behind/above & leaves ant as N of the pterygoid canal

FORAMEN MAGNUM (unpaired)

Site In occipital bone in post cranial fossa

Contains Medulla oblongata, meninges, vertebral arts, ant & post spinal arts, spinal accessory Ns (XI), sympathetic plexus on vertebral art, apical ligament of dens, tectorial membrane

FORAMEN OVALE

Site In greater wing of sphenoid bone in middle cranial fossa

Contains Mandibular N (Vc), lesser petrosal N, accessory meningeal art

FORAMEN ROTUNDUM

Site In greater wing of sphenoid bone in middle cranial fossa

Contains Maxillary N (Vb)

FORAMEN SPINOSUM

Site In greater wing of sphenoid bone in middle cranial fossa

Contains Middle meningeal vessels, meningeal br of mandibular N (Vc)

FORAMEN TRANSVERSARIUM

Site In pedicle of cervical vertebrae bordered by — lat: intertubercular lamella (costotransverse bar), med: body of vertebra

Contains Vertebral art & V in C1−6. Vein only in C7

GREATER PALATINE FORAMEN

Site Between maxilla & palatine bone at lat edge of hard palate

Contains Greater palatine N & vessels

HYPOGLOSSAL CANAL
Site In occipital bone above condyle in post cranial fossa
Contains Hypoglossal N (XII)

INCISIVE CANAL
Site In ant maxilla extending from nose to incisive foramina
Contains Nasopalatine N, greater palatine vessels

INCISIVE FORAMEN
Site Midline, in ant hard palate. Openings of incisive canals into incisive fossa
Contains Nasopalatine N, greater palatine vessels

INCISIVE FOSSA (unpaired)
Site Median, in ant hard palate leading upwards to incisive foramina
Contains Nasopalatine Ns, greater palatine vessels

INFERIOR ORBITAL FISSURE
Site Between greater wing of sphenoid bone & maxilla
Contains Infra-orbital & zygomatic brs of maxillary N (Vb), infra-orbital vessels, inf ophthalmic Vs, orbital brs of pterygo-palatine ganglion

INFRA-ORBITAL CANAL
Site Within orbital aspect of maxilla
Contains Infra-orbital N & vessels

INFRA-ORBITAL FORAMEN
Site Below infra-orbital margin in maxilla. Ant opening of infra-orbital canal
Contains Infra-orbital N & vessels

INTERNAL ACOUSTIC MEATUS
Site In post surface of petrous temporal bone in post cranial fossa
Contains Facial N (VII), nervus inter-medius, vestibulocochlear N (VIII), labyrinthine art

INTERVERTEBRAL FORAMEN
Site Between vertebrae, bordered by – sup & inf: pedicles of vertebrae, ant: vertebral body & intervertebral disc, post: lig flavum covering sup & inf articular processes
Contains Spinal art & V, dorsal root ganglion, spinal N. Nerves C1–7 emerge via foramen above same numbered vertebra; nerve C8 exits below C7 vertebra & below this all nerves emerge via foramen below the same numbered vertebra

JUGULAR FORAMEN
Site Between jugular fossa of petrous temporal bone & occipital bone in post cranial fossa
Contains Glossopharyngeal N (IX), vagus (X), accessory N (XI), inf petrosal & sigmoid sinuses enters it, int jugular V emerges below

LESSER PALATINE FORAMINA
Site Two or three foramina in med & inf aspects of pyramidal process of palatine bone
Contains Lesser palatine Ns & vessels

MANDIBULAR CANAL (INFERIOR ALVEOLAR CANAL)
Site In body & ramus of mandible between mandibular & mental foramina
Contains Inf alveolar N & vessels

MANDIBULAR FORAMEN (INFERIOR ALVEOLAR FORAMEN)
Site Med aspect of ramus of mandible, overlapped anteromedially by lingula
Contains Inf alveolar N & vessels

11

MASTOID FORAMEN
Site In petrous temporal bone in post cranial fossa, post to sigmoid groove. Exits behind mastoid process
Contains Emissary V connecting sigmoid sinus & occipital Vs, meningeal br of occipital art

MENTAL FORAMEN
Site Outer aspect of ant ramus of mandible by second premolar tooth, leading from mandibular (inf alveolar) canal
Contains Mental N & vessels

NASOLACRIMAL CANAL
Site Between lacrimal bone & maxilla at ant/inf/med corner of orbit
Contains Nasolacrimal duct

OPTIC CANAL
Site In body of sphenoid bone in middle cranial fossa between body & two roots of lesser wing
Contains Optic N (II), dural sheath, ophthalmic art

PALATOVAGINAL CANAL
Site Between upper surface of sphenoidal process of palatine bone & lower surface of vaginal process of root of med pterygoid plate in base of skull
Contains Pharyngeal Ns from maxillary (Vb) and pterygopalatine ganglion & pharyngeal br of maxillary art

PETROSQUAMOUS FISSURE
Site Between squamous temporal bone & tegmen tympani (petrous temporal bone)
Contains No structures

PETROTYMPANIC FISSURE
Site Between tympanic part (plate) of temporal bone & tegmen tympani (also part of temporal bone) in base of skull
Contains Chorda tympani, ant lig of malleus, ant tympanic br of maxillary art

PTERYGOID CANAL
Site In pterygoid process of sphenoid bone connecting ant wall of foramen lacerum to pterygopalatine fossa
Contains N & art of pterygoid canal

PTERYGOMAXILLARY FISSURE
Site Between lat pterygoid plate & post surface of maxilla connecting infra-temporal & pterygopalatine fossae, continuous above with post end of inf orbital fissure
Contains Maxillary art & N (Vb), sphenopalatine Vs

SPHENOIDAL FORAMEN
Site In greater wing of sphenoid in middle cranial fossa med to foramen ovale (40% of skulls) (venous foramen of Vesalius)
Contains Emissary V connecting cavernous sinus & pterygoid plexus

SPHENOPALATINE FORAMEN
Site Between body of sphenoid bone & sphenopalatine notch of palatine bone (sup border of perpendicular plate & orbital & sphenoidal processes). In med wall of pterygopalatine fossa
Contains Sphenopalatine art, nasopalatine & sup nasal Ns from pterygopalatine fossa

SQUAMOTYMPANIC FISSURE
Site Between tympanic part (plate) of temporal bone & mandibular fossa (squamous temporal bone) in base of skull. It is divided by tegmen tympani (petrous temporal bone) into petrotympanic and petrosquamous fissures
Contains Deep auricular br of maxillary artery

STYLOMASTOID FORAMEN
Site Between styloid & mastoid processes of temporal bone in base of skull

Contains Facial N (VII) & stylomastoid br of post auricular art

SUPERIOR ORBITAL FISSURE
Site Between body & lesser & greater wings of sphenoid bone in middle cranial fossa
Contains Ophthalmic N (Va) (lacrimal, frontal, nasociliary brs), ophthalmic Vs, oculomotor N (sup & inf divs) (III), trochlear N (IV), abducent N (VI), sympathetic fibres, brs of middle meningeal & lacrimal arts

SUPRA-ORBITAL FORAMEN
Site In supra-orbital margin of frontal bone, 2 cm from midline
Contains Supra-orbital N & vessels

VERTEBRAL FORAMEN (unpaired)
Site Bordered by — ant: body of vertebra, post: laminae, lat: pedicles & articular processes. Collectively making the spinal canal
Contains Spinal cord/cauda equina, dura, archnoid & pia mater, cerebrospinal fluid, internal vertebral venous plexus & spinal arts

VOMEROVAGINAL CANAL
Site Between lower aspect of ala of vomer & upper aspect of vaginal process of root of med pterygoid plate in base of skull (not always present)
Contains Pharyngeal br of sphenopalatine art

ZYGOMATICOFACIAL FORAMEN
Site In lat surface of zygomatic bone
Contains Zygomaticofacial N & vessels

ZYGOMATICO-ORBITAL FORAMEN
Site In orbital surface of zygomatic bone
Contains Zygomatic br of maxillary N (Vb)

ZYGOMATICOTEMPORAL FORAMEN
Site In posteromedial surface of zygomatic bone
Contains Zygomaticotemporal N & vessels

12: SPACES OTHER THAN SKULL AND SPINE

Note: (1) Including fossae, spaces, rings, canals, triangles, sacs and foramina. (2) All paired unless otherwise indicated.

12

ADDUCTOR (HUNTER'S/ SUBSARTORIAL) CANAL

Site A groove in thigh extending from apex of femoral triangle to hiatus in adductor magnus. Bordered by — lat: vastus medialis, med: adductor longus & magnus, roof: fascia in which lies the subsartorial plexus & on which lies sartorius

Contains Femoral art & vein; saphenous N; N to vastus medialis

ANTERIOR TRIANGLE OF NECK

Site Borders — inf border of mandible, midline & ant border of sternocleido-mastoid. Subdivided into carotid, digastric, submental & muscular triangles

Contains Muscles: digastric, stylohyoid, mylohyoid, geniohyoid, sternohyoid, omohyoid, thyrohyoid, sternothyroid, platysma. Hyoid bone, larynx, thyroid & parathyroid glands, trachea, oesophagus, submandibular gland, lymph nodes. Arteries: common, int & ext carotids; brs of ext carotid: sup thyroid, ascending pharyngeal, lingual, facial (submental). Mylohyoid art (maxillary via inf alveolar). Veins: int & ant jugular. Nerves: hypo-glossal, ansa cervicalis, vagus & its int, ext & recurrent laryngeal & pharyngeal brs, mylohyoid N from Vc via inf alveolar N

CUBITAL FOSSA

Site Triangular space in ant aspect of arm. Borders — sup: intercondylar line, med: lat border of pronator teres, lat: med border of brachioradialis, floor: brachialis, supinator, roof: fascia (see below for what lies in it)

Contains From med to lat: median N, brachial art & its accompanying Vs, biceps tendon, radial & post interosseous Ns seen under edge of brachioradialis. Roof: bicipital aponeurosis, median basilic & cephalic Vs, med & lat cutaneous Ns of forearm

DEEP INGUINAL RING

Site In lower abdominal wall above mid point of inguinal lig. Borders — sup & lat: curved fibres of transversus abdominis. Inf: inguinal lig. Med: transversalis fascia & inf epigastric vessels. Int spermatic fascia attached to its edges

Contains Vas deferens; testicular, vasal, cremasteric arts & Vs; obliterated processus vaginalis; genital br of genitofemoral N; autonomic Ns; lymphatics

EPIPLOIC FORAMEN OF WINSLOW (ADITUS TO LESSER SAC) (unpaired)

Site In upper abdomen. Borders — ant: portal V, post: inf vena cava, inf: first part of duodenum, sup: caudate lobe of liver

Contains Nil

FEMORAL RING & CANAL

Site In lower abdomen. Femoral ring is upper end of femoral canal. Borders — ant: inguinal lig, med: lacunar lig, post: pectineal lig & pectineus, lat: femoral V

Contains Cloquet's node; lymphatics

FEMORAL TRIANGLE

Site In thigh. Borders — med: med border of adductor longus, lat: med border of sartorius, sup: inguinal lig, floor: adductor longus, pectineus, iliacus & psoas, roof: fascia

Contains Femoral N, art, V & their brs; deep inguinal lymph nodes

GREATER SCIATIC FORAMEN

Site In pelvis between greater sciatic notch of ischium & both sacrotuberous & sacrospinous ligs

Contains From above downwards: sup gluteal N & vessels, piriformis, inf gluteal N & vessels, int pudendal art, pudendal N, sciatic N, post femoral cutaneous N, perforating cutaneous N, N to obturator internus, N to quadratus femoris

12

INGUINAL CANAL

Site In lower abdomen between deep & superficial inguinal rings. Borders — ant: external oblique abdominis & a small portion of internal oblique abdominis, post: transversalis fascia, inf epigastric vessels & conjoint tendon, sup: curved fibres of internal oblique & transversus abdominis, inf: inguinal lig

Contains Vas deferens/round lig of uterus; testicular, cremasteric & vasal arts & Vs; obliterated processus vaginalis; ilioinguinal, genital br of genitofemoral & autonomic Ns; lymphatics; int spermatic & cremasteric fasciae

INGUINAL (HASSELBACH'S) TRIANGLE

Site Post aspect of ant abdominal wall in inguinal region. Borders — lat: inf epigastric art, med: lat edge of rectus abdominis, inf: inguinal lig, floor: transversalis fascia, conjoint tendon & post wall of inguinal canal

Contains Nil. Site of direct inguinal herniation

ISCHIO-ANAL (ISCHIORECTAL) FOSSA

Site Wedge-shaped area lat to anal canal. Borders — med: anal canal & levator ani, lat: obturator internus & ischial tuberosity, inf (floor/base): post aspect of perineal body, urogenital diaphragm, sacrotuberous lig & gluteus maximus

Contains Fat; pudendal canal containing pudendal N; int pudendal vessels; inf rectal br of pudendal N

LATERAL TRIANGULAR SPACE

Site In post wall of axilla. Borders — sup/med: teres major, inf/med: long head of triceps, lat: med shaft of humerus

Contains Radial N; profunda brachii vessels

LESSER SAC (OMENTAL BURSA)

(unpaired)

Site Diverticulum from general peritoneal cavity in upper abdomen opening via epiploic foramen of Winslow

Contains Its peritoneal lining lies against — ant (from above down): post surface of liver; lesser omentum; body & fundus of stomach; greater omentum. Inf: transverse colon. Post: inf vena cava; first 2.5 cm of duodenum; aorta; coeliac trunk & brs; body of pancreas; left suprarenal gland; upper pole left kidney; sup: caudate lobe of liver; med (right): opening of sac with inf vena cava in post edge; portal V, hepatic art & bile duct in ant free edge

LESSER SCIATIC FORAMEN

Site In pelvis between lesser sciatic notch of ischium & both sacrotuberous & sacrospinous ligs

Contains Passing out: tendon of obturator internus. Passing in: N to obturator internus, int pudendal art, pudendal N

MEDIAL TRIANGULAR SPACE

Site In post wall of axilla. Borders — sup/med: subscapularis (teres minor viewed from behind), inf/lat: teres major, lat: long head of triceps

Contains Circumflex scapular art

OBTURATOR CANAL

Site In ant aspect of obturator foramen in lat wall of true pelvis. Borders — ant: post pubic ramus, sup/inf/med: obturator int & its fascia

Contains Obturator N & vessels

POPLITEAL FOSSA

Site Diamond shaped, behind knee. Borders — sup/lat: biceps femoris, sup/med: semitendinosus & semimembranosus, inf/med & lat: heads of gastrocnemius, floor: post distal femur, post capsule of knee & popliteus, roof: fascia

12

Contains Plantaris; popliteal art & V & brs; tibial, common peroneal, sural & sural communicating Ns; lymph nodes & fat. Short saphenous V & post femoral cutaneous N in fascia of roof

POSTERIOR TRIANGLE OF NECK
Site Between post border of sternocleidomastoid, ant border of trapezius & middle third of clavicle. Floor: prevertebral fascia over semispinalis capitis, splenius capitis, levator scapulae, scalenus medius & ant. Roof: investing layer of deep fascia
Contains Occipital, transverse cervical, suprascapular & third part of subclavian arts; transverse cervical, suprascapular & ext jugular Vs; muscular & cutaneous brs of cervical plexus (lesser occipital, great auricular, transverse cervical, supraclav-icular); three trunks of brachial plexus; spinal accessory N; inf belly of omohyoid; superficial cervical lymph nodes

PUDENDAL (ALCOCK'S) CANAL
Site Lies within a fascial sheath in lat wall of ischio-anal canal between lesser sciatic notch & deep perineal pouch. Borders — lat: obturator internus & ischial tuber-osity, med: fat
Contains Pudendal N; int pudendal vessels

QUADRANGULAR SPACE
Site In post wall of axilla. Borders — sup: subscapularis (teres minor viewed from behind), inf: teres major, med: long head of triceps, lat: med shaft of humerus
Contains Axillary N; post circumflex humeral art & V

SUPERFICIAL INGUINAL RING
Site In lower abdominal wall at med end of inguinal canal as a V-shaped opening of ext oblique aponeurosis. Ext spermatic fascia is attached to its edges
Contains Vas deferens; testicular, cremasteric & vasal arts & Vs; obliterated processus vaginalis; ilioinguinal, genital br of genitofemoral & autonomic Ns; lymphatics; int spermatic & cremasteric fasciae

UROGENITAL TRIANGLE
Site In perineum. Borders — lat: ischiopubic rami, ant: post aspect of symphysis pubis, post: transverse line at level of perineal body & ischial tuberosities
Contains Deep perineal pouch (space) within sup & inf fascial layers of the urogenital diaphragm perforated by urethra with surrounding ext sphincter. Deep & superficial transverse perinei; brs of int pudendal vessels; pudendal N. In male diaphragm contains bulbo-urethral glands & supports penis & scrotum from its inf surface. In female it is perforated by vagina

12

13: POSITION OF STRUCTURES ACCORDING TO VERTEBRAL LEVELS

Note: (1) A / (oblique stroke) between two levels indicates that the structure lies between these two vertebrae. (2) A − (dash) between two levels indicates that the structure occupies the equivalent to these vertebrae inclusively.

C1   Spinal accessory nerve crosses lateral mass of atlas
Open mouth and dens

C2   Superior cervical ganglion

C3   Body of hyoid bone

C4   Upper border of thyroid cartilage
Bifurcation of common carotid arteries

C6   Cricoid cartilage
Larynx becomes trachea
Pharynx becomes oesophagus
Middle cervical ganglion
Vertebral artery enters foramen transversarium of C6 vertebra
Carotid tubercle of Chassaignac
Inferior thyroid artery crosses to thyroid gland

C7   First clearly palpable spinous process (vertebra prominens)
Stellate/inferior cervical ganglion

T1   Superior border of scapula

T2/3   Suprasternal notch

T3   Medial end of spine of scapula
End of oblique fissure posteriorly

T3/4   Bifurcation of trachea
Start of arch of aorta

T3−4   Manubrium sterni

T4   End of arch of aorta
Azygos vein enters superior vena cava

T4/5   Manubriosternal angle of Louis

T5   Thoracic duct crosses midline

T5−8   Sternum

T6   Upper border of liver

T7   Inferior angle of scapula
Accessory hemiazygos vein crosses midline to azygos vein

T8   Caval opening in diaphragm
- Inferior vena cava
- Right phrenic nerve
Left phrenic nerve pierces diaphragm lat to central tendon
Hemiazygos vein crosses to right to join azygos vein

T8/9   Sternoxiphisternal joint

T9   Superior epigastic vessels traverse diaphragm
Xiphoid

T10   Oesophageal opening in diaphragm
- Oesophagus
- Left gastric vessels
- Anterior and posterior vagi

T12   Aortic opening in diaphragm
- Aorta
- Azygos & hemiazygos veins
- Thoracic duct
Origin of coeliac axis (lower border of T12)
Splanchnic nerves pierce cura of diaphragm
Sympathetic trunk passes under medial arcuate ligament

13

L1   Transpyloric plane of Addison
(half way between suprasternal
notch and pubis)
- Fundus of gallbladder
- Hila of kidneys
- Second part of duodenum
- Neck of pancreas
- Origin of superior mesenteric
  artery
- Origin of portal vein
- Pylorus
- Attachment of transverse
  mesocolon
- Hilum of spleen (spleen on ribs
  9, 10, 11)
- Tip of 9th costal cartilage

L1/2   Origin of renal arteries
Spinal cord ends in adults

L2   Subcostal plane
- Formation of azygos and
  hemiazygos veins
- Duodenojejunal flexure,

ligament of Treitz (upper border
of L2)

L3   Origin of inferior mesenteric artery

L3/4   Umbilicus

L4   Supracristal plane (iliac crests)
- Bifurcation of aorta

L5   Formation of inferior vena cava

S2   Sacral dimple
Mid point of sacroiliac joint
Post superior iliac spine
Dural sac ends

S3   Start of rectum

S4   Sacral hiatus
End of vertebral canal

Co1   Filum terminale inserts

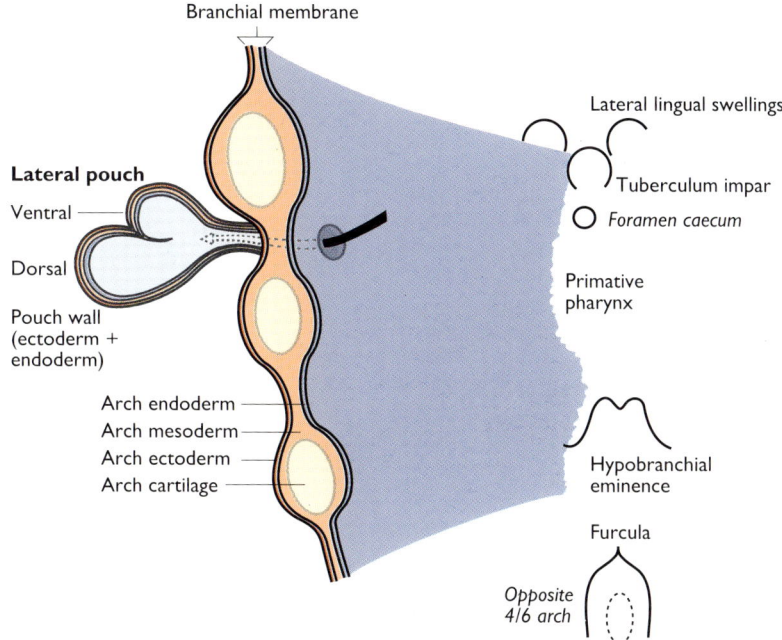

Branchial derivatives

Branchial derivatives

Arch	Arch derivatives			Lateral pouch derivatives		Artery	Nerve
	Cartilages, bones, ligaments	Muscles	Endoderm	Endoderm	Ectoderm		
	Mesoderm						
1 'Mandibular' arch cartilage = Meckel's	Incus Malleus Anterior ligament of malleus Sphenomandibular ligament (Lingula) (Mandible*)	Masseter Temporalis Pterygoids Mylohyoid Anterior belly of digastric Tensor veli palatini Tensor tympani	Mucous membrane and glands of anterior 2/3 of tongue	Auditory tube Inner layer of tympanic membrane (Part of middle ear) (Mastoid antrum)	External acoustic meatus Outer layer of tympanic membrane (Tragus of ear) (Skin of lower face)	Part of maxillary artery	Mandibular division of trigeminal (Vc)
2 'Hyoid' arch cartilage = Reichert's	Upper body and lesser cornu of hyoid Stylohyoid ligament Styloid process Stapes	Stapedius Stylohyoid Posterior belly of digastric Muscles of facial expression including buccinator and platysma	—	Supratonsillar fossa Tonsillar crypts Surface epithelium of tonsil** (Part of middle ear)	Overgrowth of ectoderm over arches 3, 4 & 6	Stapedial artery	Facial (VII)

14

3 'Thyrohyoid' arch	Inferior body and greater cornu of hyoid	Stylopharyngeus	Mucous membrane and glands of posterior 1/3 of tongue, valleculae and anterior epiglottis	Ventral: epithelial cells of thymus** Dorsal: inferior parathyroid	—	Internal carotid artery (including carotid sinus)	Glossopharyngeal (IX)
4	Thyroid cartilage	Palatoglossus Palatopharyngeus Salpingopharyngeus Cricothyroid Levator veli palatini Striated of oesophagus Pharyngeal constrictors	—	Ventral: ultimobranchial bodies† Dorsal: superior parathyroid	—	Right: part of right subclavian artery Left: aortic arch	Vagus (X) Pharyngeal and superior laryngeal branches
6	Cricoid cartilage Vocal ligaments Arytenoid, corniculate and cuneiform cartilages	Cricopharyngeus All intrinsic muscles of larynx	—	Lung buds	—	Ventral: pulmonary artery Dorsal: ductus arteriosus	Vagus (X) Recurrent laryngeal branch

* The mandible forms in membrane around the ventral aspect of the first arch cartilage.

** The lymphoid tissue of the tonsil and thymus arises from the surrounding mesenchyme and is not arch derivative.

† Ultimobranchial bodies develop from ventral parts of fourth (and possibly fifth) pouch and fuse with the developing thyroid to give parafollicular (C) cells which produce calcitonin.

Note: (1) The thyroid gland arises from between the first and second arch as a diverticulum (thyroglossal duct) which grows downwards leaving the foramen caecum at its origin. (2) The epiglottis comes from the inferior part of the hypobranchial eminence and is thus not a true arch derivative. (3) Bracketed information is of additional interest

14